T0275655

Elements in Philosophy of Science
edited by
Jacob Stegenga
University of Cambridge

FEMINIST PHILOSOPHY OF SCIENCE

Anke Bueter
Aarhus University

Shaftesbury Road, Cambridge CB2 8EA, United Kingdom

One Liberty Plaza, 20th Floor, New York, NY 10006, USA

477 Williamstown Road, Port Melbourne, VIC 3207, Australia

314–321, 3rd Floor, Plot 3, Splendor Forum, Jasola District Centre, New Delhi – 110025, India

103 Penang Road, #05–06/07, Visioncrest Commercial, Singapore 238467

Cambridge University Press is part of Cambridge University Press & Assessment, a department of the University of Cambridge.

We share the University's mission to contribute to society through the pursuit of education, learning and research at the highest international levels of excellence.

www.cambridge.org
Information on this title: www.cambridge.org/9781009507257

DOI: 10.1017/9781108956055

When citing this work, please include a reference to the DOI 10.1017/9781108956055

First published 2024

A catalogue record for this publication is available from the British Library.

ISBN 978-1-009-50725-7 Hardback
ISBN 978-1-108-95834-9 Paperback
ISSN 2517-7273 (online)
ISSN 2517-7265 (print)

Feminist Philosophy of Science

Elements in Philosophy of Science

DOI: 10.1017/9781108956055
First published online: June 2024

Anke Bueter
Aarhus University

Author for correspondence: Anke Bueter, abueter@cas.au.dk

Abstract: Feminist scholars have identified pervasive gender discrimination in science as an institution, as well as gender bias in the very content of many scientific theories. An ameliorative project at heart, feminist philosophy of science has inquired into the social and epistemological roots and consequences of these problems and into their potential solutions. Most feminist philosophers agree on a need for diversity in scientific communities to counter the detrimental effects of gender bias. Diversity could thus serve as a unifying concept for a potential consensus of the field. Yet there are substantial differences in the kinds and roles of diversity envisaged. This element argues that we need diversity, both in terms of social locations and of values, to overcome former biases and blind spots. Diversity as such, however, is insufficient. To reap its epistemic benefits, diversity also needs to be institutionalised in a way that counters various forms of epistemic injustice.

Keywords: gender bias in science, science and values, feminist standpoint theory, critical contextual empiricism, diversity

ISBNs: 9781009507257 (HB), 9781108958349 (PB), 9781108956055 (OC)
ISSNs: 2517-7273 (online), 2517-7265 (print)

Contents

1 Introduction

Feminist philosophy of science is concerned with matters of gender (in)equality and their impacts on scientific knowledge, as well as with the consequences of gender bias in research for how we think about and treat people of different genders. Prima facie, combining feminism (a political position) and philosophy of science (an epistemological and metaphysical inquiry) might appear misguided at worst, or at best only marginally relevant. It may seem misguided because it mixes matters of ideology and science in an epistemically and politically irresponsible fashion. Especially in its early days, feminist philosophy of science was often accused of replacing central scientific ideals of objectivity, truth, and methodological rigour with dogmatic or wishful thinking. The very idea of a feminist (philosophy of) science, it was argued, undermines the epistemic trustworthiness of (philosophy of) science as well as feminism. Interpreted in a slightly more charitable way, feminist philosophy of science might be seen as a project that reflects upon areas of science directly relevant to matters of gender/sex (e.g., purported cognitive difference between the sexes) or of particular interest to women (e.g., women's health). Understood in this latter way, feminist philosophy of science makes additions to the discipline that focus on specific subject areas. For example, this could mean to uncover gender bias in former research in these fields.

Such work has been very important, but feminist philosophy of science is, in fact, a far more ambitious endeavour. While it started from empirical case studies in areas directly relevant to gender/sex, this has led to a thorough rethinking of evidential standards and key concepts such as confirmation, objectivity, and value-freedom in science more generally. The relevance of feminist philosophy of science is thus by no means limited to feminists. Overall, feminist philosophy of science has been a very important driver towards an understanding of science as an essentially social enterprise. Importantly, this does *not* mean collapsing the rational into the social or denying science its traditional claim to epistemic authority. At the same time, science is regarded not only as social but also as deeply political. Feminist philosophers of science are particularly interested in the effects of power relations on scientific knowledge and knowers. They thus focus on categories that are related to the distribution of social power; starting with gender, but also extending to variables such as race, class, or (dis)ability, as well as their interactions.

Feminist philosophy of science is, accordingly, often critical of specific areas of scientific research, as well as of its more abstract philosophical underpinnings and socio-political consequences. Nevertheless, it is, at heart, an ameliorative project. A central assumption in most of contemporary feminist philosophy of science is that feminism, science, and its philosophy are potential allies, not enemies. (Philosophy of) science, at its best, can help the feminist cause; for example, by providing well-founded relevant knowledge. Feminism, in turn, can further science and the philosophy of science; for example, by correcting past mistakes, by deepening our understanding of what 'good science' means, or simply by providing new perspectives and questions.

Over the last decades, feminist philosophy of science has, accordingly, stimulated rich debates on what science is and should be like. It will be hard to do them all justice within the short span of this Element. I aim to provide an overview of the central lines of the field's historical development, leading up to more current discussions and some open questions. Section 2 begins with a brief overview of issues related to the discrimination against female scientists and its connection to gendered ideas about rational knowers. Section 3 traces the impact of gender bias on the contents of scientific theories by recourse to empirical case studies, introducing infamous examples from the fields of evolutionary biology, primatology, and medicine. Section 4 sketches different philosophical responses such as feminist standpoint theory and feminist empiricism, and discusses the (in)adequacy of early critiques of the developing field. Section 5 focuses on the question of values in science, which has long been a pivotal concern in the field. Many of the arguments here point in the direction of value-freedom being unattainable, and potentially even misguided as a normative ideal. Instead of value-freedom, feminists of different theoretical orientations have tended to focus on diversity in the scientific community as the best way to ensure that values in science are not epistemically detrimental but potentially fruitful.

Section 6 therefore discusses whether diversity can serve as the central, unifying idea for a consensus position in the field. Yet, the devil lies in the details here. While, for instance, feminist standpoint theorists and critical contextual empiricists agree that diversity is crucial, they disagree over what kind of diversity is important and why. I will advocate an approach that conceptualises diversity in terms of different values as well as social locations and assigns them a causal role in identifying and overcoming biases and blind spots. This contrasts with standpoint theoretical accounts that assign values the role of good reasons in themselves. Standpoint theorists are correct, however, in arguing that not all perspectives are on a par epistemologically. This is so because those perspectives that have been historically excluded from science will be most likely to challenge existing assumptions and priorities that lead to bias and systematic ignorance. I will connect this to recent discussions of epistemic injustice in science, and I will argue that a diverse scientific community is epistemically advantageous only in so far as it is also epistemically just.

1.1 A Note on Terminology

Before I get started, some remarks on the gender-related concepts used in this Element. The distinction between sex as a biological category and gender as a social one has been very influential and helpful in feminist thought, as it served to challenge automatic assumptions of biological determinism (in the sense of feminine or masculine traits and behaviours being caused by one's biological sex). Some scholars nowadays use the terms 'sex/gender' or 'gender/sex' to indicate that these dimensions are often interrelated, albeit not in a deterministic manner. For example, some bodily features are impacted by gendered ideas about what women/men are like, which affect variables such as nutrition, kinds of work, or types of exercise (Clune-Taylor 2020). The debate on these concepts is also connected to various metaphysical and political questions, which I have to bracket here (for an overview, see Mikkola 2023). In the following, I will use 'sex' when referring to primarily biological and 'gender' when referring to primarily social aspects. Where they seem interrelated or it is unclear which level is prioritised, I will use the combined term 'gender/sex'.

A related problem is how to name the different sexes/genders. I will often use the terms 'woman' and 'man', simply because they have been used extensively both in the history of science and of (feminist) philosophy of science (and congruence here will enhance readability). Often this was done without an explication of these concepts because it is presumed that people are of either female or male sex, and that they also identify themselves, and are identified by others, as being of female or male gender. This presumption runs into numerous difficulties, however. For instance, people of female sex have long been discriminated against in science because of gendered prejudices against women, whether or not they self-identify as female. Moreover, using these categories in a binary fashion ignores all kinds of non-binary people and people who do not identify with the sex/gender assigned to them. Importantly, gendered oppression affects not just biological females, but basically everybody who is not a cisgendered male whose gender identity matches the sex he was assigned at birth and who is read as such by other members of society. An umbrella term might be 'non-cisgender-male', yet it seems unfortunate to identify those suffering from oppression negatively and via reference to a male standard. In what follows, I will therefore use the terms woman/man and female/male somewhat loosely to refer to individuals with the respective gender identity, and I will by no means presume that this is a binary opposition. Sometimes these terms will also be used to refer to the biological sexes, especially when reconstructing episodes from the history of science.

2 The Gender of Scientific Knowers

There are obvious connections between feminism and science, both on the structural, institutional level, and on the level of content. On the one hand, women have been discriminated against as subjects of scientific knowledge and have been excluded from scientific careers. On the other hand, science has often been instrumental in gendered oppression by ways of making claims about women's supposed nature, such as women being less intelligent, less rational, more emotional, more caring, more passive, and more submissive than men. The presumed upshot is that women are better suited to caretaking and service to others than to leading a life of the mind. Rather than an issue of discrimination and unfairness, women's underrepresentation in academia would thus be an expression of the natural order of things. This, of course, contributes to and perpetuates the underrepresentation of women in scientific careers, cementing the view that women lack scientific interests and abilities further in a vicious cycle. As we will see below, a lack of women in the scientific community also means a lack of critical perspectives on the very theories and data that were supposed to support the relevant assumptions about women's nature. Exclusion and discrimination on the institutional level and gender bias in scientific content thus go hand in hand and reinforce each other. In what follows, I will look at these two levels in some more detail.

2.1 Issues of Equality and Representation

Women have been formally denied access to higher education and scientific careers for centuries. Even today, women are still underrepresented in many disciplines, and face a myriad of obstacles as researchers. Denmark, for example, is above the European average according to the gender equality index of the European Union. The data show that by now, more women than men graduate in tertiary education. Options for childcare are also comparatively good. Still, only 23 per cent of all full professors in 2019 were female. In addition, there is a high degree of segregation regarding the fields women versus men graduate in (with 53 per cent versus 27 per cent being found in education, health and welfare, humanities, and arts). In 2016, the percentage of female professors in the natural sciences was 12 per cent, and in the technological sciences only 8 per cent. Women also work under precarious conditions more often than men in the higher education sector (e.g., with part-time or fixed-term contracts). Looking at the wider field of R&D, only 10 per cent of patent applications are made by women.[1] While there are of course important differences between various countries, EU-wide data show similar patterns:

[1] Cf. European Institute for Gender Equality (2021). *Gender Equality Index,* https://eige.europa.eu/gender-equality-index/2021/domain/knowledge/DK; Danish Society for Women in Science (2023). *Hiring Statistics*, https://danwise.org/facts-and-statistics/hiring-statistics/; Styrelsen for

Whereas more than half of all students in Europe were female in 2018, only about 26 per cent of full professors were. Zooming in on the STEM disciplines, these numbers reduced to 32 per cent of female students and 19 per cent of female full professors (European Commission 2021, ch. 6). Gender segregation also pervades science and engineering in the USA; for example, in 2021, fewer than one-third of doctoral degrees were awarded to women in engineering, mathematics and statistics, and computer and information sciences, whereas about half of doctoral degrees in the social and biomedical sciences and almost three quarters in psychology were obtained by women (NSB 2023).

While it is clear that the gender distribution is still skewed, especially in the prestigious STEM disciplines, there is considerable debate regarding the causes of female underrepresentation in science (and philosophy). Most feminists would hold that this has to do with gender bias and covert or overt discrimination. Alternative explanations contend that this is due not to a lack of equal opportunities but to self-selection, with women deciding for other career paths or choosing to drop out of science, especially its most prestigious fields (e.g., Ceci & Williams 2011; Cole 1987; for a critique, cf. Rolin 2006).[2] Others have proposed that the lack of women in science and technology results from women's lack of aptitude (e.g., innate mathematical ability) and/or effort (e.g., the willingness to put in eighty hours per week) (Summers 2005; for a critical response, cf. Handelsman et al. 2005).[3]

However, there is by now a huge body of literature documenting various subtle (and not so subtle) mechanisms of exclusion and discrimination that questions these alternative explanations (cf. also Crasnow 2020). It has been shown that gender inequality can arise not only from formal barriers to access, overt discrimination, or sexual harassment in the workplace but also from less visible mechanisms. These range from factors such as unconscious perpetuation of gendered stereotypes by parents, field-specific ability beliefs,[4] a lack of role models and a sense of belonging in education, to gendered distributions of

Forskning og Uddannelse (ed.) (2020). Mænd og kvinder på de danske universiteter: Danmarks talentbarometer 2019, https://ufm.dk/publikationer/2020/filer/talentbarometer-2019.pdf.

[2] To be fair, current researchers mostly agree that the causes of female underrepresentation in STEM fields are multi-factorial. For instance, Ceci and Williams (2011) acknowledge that the choice to drop out of a scientific career, or never to pursue one to begin with, can be free or 'constrained', e.g., by biological factors such as women's restricted span of fertility (in combination with the time pressures of the academic labour market). Clearly, such choices do not occur in a social vacuum. It is thus often a question of what factors are emphasised (and which might be downplayed), or of how their interrelation is conceptualised. For a review of different proposed causal factors and the respective evidence, cf. Wang and Degol (2017).

[3] For critical discussions of research on cognitive differences between sexes, cf., e.g., Bluhm (2020), Crasnow (2020), and Kourany (2016).

[4] Common beliefs that certain fields such as mathematics or philosophy require innate talent or brilliance (versus effort and learning) correlate with lower numbers of women in such fields. Arguably, this is so because brilliance is stereotypically associated with white men (Leslie et al. 2015).

teaching and administrative duties, implicit bias in evaluation processes, lack of access to informal networks, and gendered citation patterns (for an overview, see, e.g., Sugimoto & Larivière 2023; Wang & Degol 2017). Already the daily experience of micro-inequities (such as being interrupted more frequently), as has been argued early on, can have a cumulative effect that can be very consequential and overall creates a chilly climate for female academics (e.g., Hall & Sandler 1982, 1984; MIT 1999).

Over the last decades, we could witness progress in terms of understanding how gender bias affects academic careers, even though the resulting picture is generally complex and at times puzzling. We have also seen some progress in terms of the employment rates of female researchers, even though there is still a significant gender gap. In recent years, some scholars have claimed that discrimination is now behind us. For example, Ceci and Williams hold that the academy, even in the STEM disciplines, is either gender-neutral or privileges women (e.g., Ceci and Williams 2011; Williams and Ceci 2015). Yet their research suffers from a selective focus on particular points in the span of an academic career (such as appointments to tenured positions). This overlooks the myriad of other points that impact scientific careers, as well as the evidence on covert discrimination mentioned above. For example, it does not follow from the fact that appointment committees do not display gender bias in their evaluation of tenure-track candidates that everyone has an equal opportunity to even reach this point. The playing field is not level to begin with. Leuschner and Fernández Pinto (2021) provide a detailed critique of these claims and point out such issues of overgeneralisation:

> From the fact that there is no disparity in outcomes disadvantaging women in one specific academic context (e.g., manuscript or grant proposal assessment), it cannot be concluded that women do not experience discrimination in *any* academic context (Leuschner and Fernández Pinto 2021, 579).

As an example, bias against female authors can be countered by double-blind or triple-blind peer review formats (Ceci and Williams 2011; Lee et al. 2013). Yet the submission of a manuscript of a certain quality to a prestigious journal might be easier to achieve given a series of preconditions: One needs a position that allows sufficient time for research (or to have time for it after official working hours), one might profit from colleagues' constructive feedback before submission, one might be in need of co-authors with particular expertise or require access to particular technology, and one needs sufficient epistemic self-trust. All of these factors can be affected by gender and other variables related to societal power structures.

If a paper is submitted and accepted, the next question is whether it receives uptake, that is, is read and cited in the field. In recent years, bibliometric research has demonstrated extensive citation imbalances in a variety of disciplines (e.g., neuroscience: Dworkin et al. 2020; medicine: Chatterjee and Werner 2021; for a general analysis, see Sugimoto & Larivière 2023). Published work by female authors is systematically under-cited compared to that by their male counterparts. Unsurprisingly, the same pattern can be found regarding citations of papers by white authors versus authors of colour (with black women scoring worst; Kwon 2022). At the same time, citation counts and bibliometric markers such as a researcher's h-index play an increasing role in scientific careers, influencing, for example, hiring and funding decisions. Biased citation patterns therefore bias decisions made further down the line.[5]

This is rendered invisible if one just compares the qualifications and successes of applicants in a synchronic, decontextualised, and politically insensitive manner. Let's say we have two top candidates with the same level of publications, citations, and excellent teaching evaluations. Choosing between those two (one white male and one black female) might not in itself be affected by any explicit or implicit bias – but this does not make the decision neutral if the very criteria of comparison are affected by social bias. Thus, even if some particular aspects of some of the decisions made in relation to academic careers have become more equitable over the last decades (not least thanks to the work of feminist scholars), it seems highly implausible that discrimination and social bias are now behind us, or that the underrepresentation of female scientists is not causally related to such biases.

What should, in any case, be uncontroversial is that there is a long history of exclusion of women and other marginalised groups from science. This exclusion usually worked not only via denying women access to education and careers as researchers but also went hand in hand with the creation and perpetuation of gender stereotypes by contemporary scientists. We will look at how such stereotypes found their way into scientific ideas and theories in the following sections, starting with how they shaped the very idea of a (scientific) knower.

2.2 Issues of Metaphor and Dichotomy

The history of science and its reflection in Western philosophy has been thoroughly gendered from the very beginning. To start with, the very conception of a scientific knower has not been gender-neutral: being a scientist (or philosopher)

[5] The same pattern has been demonstrated in recent studies on bias in students' evaluations of teaching. Some studies show that white men receive significantly better assessments than all other people, even though these assessments do not correlate with students' learning outcomes. Other studies show a more complex picture of 'gender (stereotype) affinity bias': female (male) students evaluate female (male) teachers better *if* these accord with stereotypical gender role behaviour, such as female teachers being more caring (e.g., Kreitzer and Sweet-Cushman 2021; Mengel et al. 2019). Down the line, this, too, can skew decisions on tenure or payment, affects self-confidence, and so on.

was conceived of as a way of living and thinking suited to male rather than female humans. As Genevieve Lloyd (1984) has put it, it is the 'man of reason' who strives to reveal nature's secrets and to (potentially) control her. Lloyd shows how reason and rationality have been portrayed as typically male domains from ancient Greece through medieval philosophy to the scientific revolution, the enlightenment, and beyond. Particularly noteworthy here is her analysis of Francis Bacon. Bacon was a seventeenth-century philosopher whose empiricist and inductivist stance was highly influential for the development of our modern notion of science. He used gendered, sexualised metaphors abundantly throughout his work. Infamously, he characterised inquiry by comparing it to a heterosexual marriage: 'Let us establish a chaste and lawful marriage between mind and nature' (cited after Lloyd 1984, 11). Mind, reason, and science are portrayed as male, whereas nature represents the female part in this marriage. The contemporary model of this marriage was, moreover, one of dominance and submission: 'Nature betrays her secrets more fully when in the grip and under the pressure of art than when in enjoyment of her natural liberty' (cited after Lloyd 1984, 11 f.).

Other feminists have also argued that the Western tradition of thought, including the history of science, has been implicitly but deeply structured by such gendered metaphors, which juxtapose masculinity with reason, rationality, and science, while characterising nature, emotion, and irrationality as feminine. The *man* of reason is detached, analytic, and rational, whereas femininity has long been conceived of in a contrasting way as intuitive, warm, emphatic, caring, holistic, corporeal, and often irrational.

While some feminist interpretations of Bacon's work have been met with criticism,[6] it seems undeniable that historically, science has not only been dominated by affluent white men and excluded other kinds of people, but also often has been thought of in a way that aligns with stereotypes of masculinity and femininity. This, in turn, has been argued to be highly consequential for the representation and success of women in scientific careers: Their underrepresentation is embedded in a certain way of thinking about thinking. In addition, this way of understanding reason and rationality in terms of 'purity' from bodily, emotional, or other worldly matters leaves us with distorted ideas about the human mind and knowledge, as Rooney (1991) argues.

[6] Some feminists have made more radical and controversial claims about how Bacon's sexualised metaphors give rise to an exploitative relation to nature, ultimately contributing to the subordination of women as well as the modern ecological crisis (Merchant 1980). Others have criticised these accounts for cherry-picking from or misunderstanding Bacon's writings, or have doubted the postulated impact of gendered metaphors (e.g., Soble 1995).

> Our history has given us what, at best, can only be described as a very impoverished discourse. [. . .]. Just as we have at best a caricature of reason, we also are left with a caricature of feeling, feeling robbed of any claim to rationality and understanding. [. . .] The history of reason, emotion, and imagination has surely been a curious one. Despite reason's articulated stance of separation from emotion and imagination, it has embedded itself in an emotional and imaginative substructure characterised largely by fear of, or aversion to, the 'feminine' (Rooney 1991, 97 f.).

The history of our concepts of reason, rationality, and by extension science has been characterised by dichotomies of reason versus emotion, abstraction versus connection, and mind versus body, that were connected to ideas about masculinity and femininity and that usually involved a devaluation of the latter, feminine part of the opposition. These dichotomies have deeply affected ethical, metaphysical, and epistemological traditions by positing unencumbered, abstract thinking as ideal. In a recent essay, Martha Nussbaum traces this ideal back to a fear of the body and, ultimately, of death, which led to the conception of an immortal, incorporeal soul in Ancient Greece: 'For Plato, the incorporeal was intelligent, lofty, lovable, and pure; and the body was stupid, base, disgusting, and impure, a prison for the soul' (Nussbaum 2022). The separation of body and soul was epitomised by René Descartes in the seventeenth century, casting a long shadow on metaphysics and epistemology. Moreover, as Nussbaum argues, the fear and disgust of the body underlying this dichotomy are differential in their application:

> In all known societies, with or without Plato, there is a second level, what I call projective disgust, in which properties of disgust are projected onto a social group that is stereotyped as the animal in opposition to the dominant group's pure soulness. They are said to be dirty, to smell bad. [. . .] Sometimes the subordinated group is a racial minority, sometimes a 'deviant' sexual group, sometimes people with disabilities, sometimes aging people, sometimes just women, who always seem to represent the body to aspiring males by contrast to the intellect and the spirit (Nussbaum 2022).

Such disgust and contempt also extend to non-human animals, which are devaluated for their presumed lack of consciousness and rationality. Some feminist thinkers have reacted to the presence of these gendered dichotomies by revaluing the traditionally scorned side. In ethics, Gilligan (1982) has argued that the dominant approach to moral issues in terms of abstract reasoning about justice and rights prioritises male over female perspectives, as the latter would approach moral issues in terms of caring, interconnectedness, and responsibility. Regarding epistemology and philosophy of science, a revaluing of 'feminine' cognitive styles has also been proposed. These cognitive styles – that is, 'women's ways of knowing' – supposedly focus on contextualisation, qualitative methods, holistic accounts, and the like (e.g., Belenky et al. 1986).

While it is problematic if certain research areas or methods are denigrated based on their being perceived as feminine (cf. Fox Keller 1983), these reappraisal approaches have met with substantial criticisms in the feminist community and beyond. They risk buying into, and therefore perpetuating, stereotypes about men versus women and their presumedly different ways of thinking, stereotypes that have played detrimental roles regarding gender equality and scientific progress. This may be connected to biological essentialism, in which the female and the feminine are identified. For example, assumptions about feminine/masculine cognitive styles might be based on hypotheses about male versus female brains functioning differently, which would make it 'natural' for men to think abstractly and for women to care about context. Yet this is not a necessary connection; one could hold that men and women have different cognitive styles based on their socialisation and social context. However, many critics have argued that even a non-essentialist insistence on different cognitive styles that aims at a revaluation of the feminine styles preserves the described dichotomies between what is symbolically gendered as masculine versus feminine. Despite the feminist intention, 'reclaiming the feminine' ultimately subscribes to stereotypes about female thinking being more concrete, worldly, emotional, caring, intuitive, holistic, and so on. By leaving the metaphorical landscape intact, such approaches fail to provide pathways to less distorted accounts of reason and knowledge (Rooney 1991).

Furthermore, many of these assumedly feminine features have been instrumental in women's oppression. As women have for so long been encouraged to cultivate traits such as sensitivity, gentleness, or humility, these traits have become 'marks of acquiescence in powerlessness' (Code 1991, 17). Reappraising only the devalued side of the described dichotomies, while leaving the dichotomies and their symbolical gendering untouched, means that such approaches remain toothless with regard to political change as well.

To summarise, our very conceptions of knowers and of science have a certain socio-cultural history and carry certain connotations. As universal and pure as they may seem, they are embedded in a social context and therefore impacted by, and themselves impacting on, power relations. This central insight has provided one of the starting points of feminist philosophy of science. Traditionally, epistemology and philosophy of science have revolved around an atomistic model of knowers, a model that considers knowers to be interchangeable, self-sufficient individuals. By contrast, '[t]he central concept of feminist epistemology is of situated knowledge: knowledge that reflects the particular perspectives of the knower' (Anderson 2020). Gaining knowledge about certain phenomena often depends on one's own situation in relation to the respective phenomena, which may be impacted by one's psychological and social situation. For example, it can be affected by one's location in the social hierarchy and interrelations with other knowers (and thereby one's gender, race, class, etc.), one's background beliefs, training, and so on. Situated knowledge is also embodied knowledge, diverging from the ideal of 'pure', incorporeal rationality.

Given an understanding of knowers as situated and interconnected in various ways, new questions come into view. Mainstream epistemology has long focused on the conditions under which an interchangeable subject S knows that p – that is, focusing on the 'knows that' part and related conditions of justification. Yet S never has been truly interchangeable but rather appeared to be so only because of the invisibility of the role of privilege and power in knowledge. As Lorraine Code argues:

> The S who could count as a model, paradigmatic knower has most commonly – if always tacitly – been an adult (but not old), white, reasonably affluent (latterly middle-class) educated man of status, property, and publicly acceptable accomplishments (Code 1991, 8).

Feminist epistemologists expand their investigation to include the impact of who S is and the context the knowledge relation is embedded in – questions that from a mainstream, individualist perspective seemed irrelevant (cf. Code 1991, 4 ff.). Traditional philosophy of science likewise focused on issues of the scientific method, which was at least in principle and implicitly considered to be implementable by interchangeable individual scientists. Current feminist epistemology of science, by contrast, focuses on the fact that research nowadays is often done by groups of people, and always within a certain societal and institutional context. It treats scientists as socially differentiated knowers who bring different backgrounds and perspectives to the research process. Such a diversity of perspectives is said to be essential to the scientific process.[7] This brings to the fore questions about relations between scientists, such as the role of trust and testimony in science, and how these relations are impacted by power distributions – and thus

[7] Feminist epistemology of science is thus understood here as a social epistemology. In general, philosophy of science has increasingly turned towards an understanding of science as a social process, starting with the demise of logical empiricism as a consequence of the historical turn introduced by Kuhn and others, as well as in response to systemic arguments such as the problem of underdetermination (see also Section 5.2). For reasons of space, it is impossible for me to sketch this whole development here (for more information, cf., e.g., Bueter 2012; Godfrey-Smith 2009; Grasswick 2018). Feminist epistemology has been central to the development of social epistemology of science, albeit its impact often remains underacknowledged by mainstream approaches (Rooney 2011). It is distinguished from social epistemology by a more explicitly critical and political focus on the general role of power relations in science, as well as on the specific variable of gender. For more on the history and impact of feminist philosophy of science, cf. Kourany (2010), Richardson (2010), and Rooney (2012).

by variables such as gender, ethnicity, class, sexual orientation, or (dis)ability. These systematic questions will be discussed in more detail in the next chapters, after I have set the stage by giving a general account of, as well as some prominent examples of, gender bias in research.

3 Gender Bias in Research

3.1 A Taxonomy of Gender Bias in Research

Over the last fifty years, work in feminist science studies (including historical, sociological, and philosophical approaches) has identified many instances of gender bias in a variety of disciplines; that is, of gender impacting research in a way that is epistemically problematic. In a wide sense, to call some piece of research biased means that we have good reasons to believe that this research could have been done systematically better (Bueter 2022). In a narrow sense, 'bias' in science is often defined in terms of the tendency to err in a certain direction, that is, as a systematic deviation from the truth. Bias, accordingly, is 'any process at any stage of inference which tends to produce results or conclusions that differ systematically from the truth' (Sackett 1979, 60).

'Gender bias' in science, as understood here, consists in (contributions to) systematic errors as well as systematic omissions in research introduced by, for example, stereotypical beliefs and prejudices. Bias can affect or result from all stages of research, including

» the choice of research topics
» the framing of hypotheses
» the use of certain concepts and definitions
» the operationalisation of concepts
» the choice of relevant versus irrelevant variables
» the choice of methods
» the choice of measurement instruments
» study design
» the implementation of methods
» data analysis
» data interpretation
» evidentiary standards
» inferential conclusions
» the dissemination of results.

In her early, groundbreaking work on non-sexist research methods, Margrit Eichler (1988) distinguished between the following major ways in which sexist dispositions can lead to biased research: androcentricity, overgeneralisation, overspecificity, gender insensitivity, and double standards. These forms of gender bias often overlap and can lead to more specific forms of bias, but they are helpful to create a general map of the terrain.

Androcentricity occurs when research focuses on the male perspective and male activities while disregarding female (or other non-male) ones. For example, this could mean writing the history of technology concentrating disproportionately on inventions made by men, while overlooking contributions from women or ignoring differential consequences of technological developments for different genders. This often goes hand in hand with certain definitions of concepts: for example, understanding 'technology' in terms of artefacts (e.g., weapons) rather than techniques (e.g., preservation of food), or in terms of 'hard' versus 'soft' artefacts (e.g., spaceships versus textiles) (Dusek 2006, ch. 9; Marçal 2021, ch. 3). Androcentricity will often produce erroneous results because it leads to conclusions that are drawn based on data characterised by systematic omissions. The same holds for *gynocentricity* as the other side of this coin. It should also be noticed that there is no uniform 'male' or 'female' perspective; rather, ideas about what counts as male versus female will often already be based on stereotypes (e.g., focusing on hunting activities in paleoanthropology because hunting is *perceived* as a typically male activity).

Overgeneralisation occurs when research generates data on only one sex or gender, but draws general conclusions from these data. Famous examples include medical trials on the efficacy of treatments for non-reproductive health issues that were done on exclusively male subjects (human as well as animal) for a long time, yet the results were assumed to apply to humans in general. In many instances, such assumptions have turned out to be false, resulting in women being treated with medication that is less effective or has more adverse effects for them, is prescribed at overdose levels for their average weight and metabolism, and so on. As Eichler notes, androcentricity and overgeneralisation often come together (as in the example of male study populations in medicine), but this is not necessarily so. First, one might also overgeneralise gynocentric results (she gives the example of drawing conclusions on the behaviour of 'parents' based on observations of mothers). Second, one might produce androcentric results without ever generalising from them (Eichler 1988, 6). If one overgeneralises results that hold for only a part of the population, one will produce biased results that systematically deviate from the truth for at least a significant part of that overall population.

Overspecificity is the flip side of overgeneralisation; that is, differentiating between genders where gender is in fact irrelevant. For Eichler, this mainly occurs on the linguistic level; for example, when characterising members of certain prestigious professions (such as doctors or scientists) as male, or referring to men only when one talks about all of humanity. This, too, usually relies on gender stereotypes and may be combined with androcentricity, such as when human interests are reduced to male interests. This phenomenon also often shows up in the use of gendered metaphors for the description of entities that are gender-neutral per se, such as nature, Earth, reason, emotion, germ cells, or bacteria (e.g., Bivins 2000; Lloyd 1984; Martin 1991).

Gender insensitivity occurs when gender as a socially significant variable is ignored, even though it is relevant. For example, this might occur when evaluating male versus female scientific aptitude based solely on the number of male versus female scientists in tenured positions, ignoring their different social realities and the social factors that make scientific careers harder to access for women in the first place. Gender insensitivity is related to overgeneralisation, but it may also occur where one has a mixed study population but fails to provide a gender-specific analysis of the study results. Usually, this will be due to a lack of awareness of the role of gender in the respective context. Where gender does play an important role, it will thus lead to biased results by ignoring important causal factors. Gender insensitivity can also be connected to biological determinism when focusing on biological differences while ignoring their social underpinnings.

Double standards apply where identical situations or behaviours are treated differently based on gender. For example, this may mean that aggression in a man is evaluated as a normal response to a certain situation, while in a woman it is interpreted as a sign of irrationality, of hormonal imbalances, or as symptomatic of mental disorder. It is related to the problem of overspecificity and may consist in the use of different methods or concepts when researching female versus male study populations, or in differential inferences from similar data. Double standards can sometimes be warranted: that is, when biological or social sex/gender variables are known to be relevant, there can be reasons for treating male and female populations differently. In such cases, resisting double standards may collapse into gender insensitivity.[8]

[8] In a more recent version of this taxonomy of bias, Burke and Eichler (2006) consider double standards as problematic when they reinforce the subordinate status of non-dominant groups. In general, their new work connects bias to the maintenance of social hierarchies as opposed to contributing to greater equity. As this presumes the legitimacy of values such as equity as integral elements of research (which will be discussed in Section 5), I stick here to Eichler's earlier framework, which defines bias in mainly epistemic terms. See also ChoGlueck (2022) for an interesting discussion of the legitimacy of double standards in trials on hormonal contraceptives for male versus female populations.

3.2 Gender Bias in Different Disciplines

Having set up this rough taxonomy, the next section will present several well-known examples of gender bias in research. As this empirical work shows, gendered background assumptions have played a pervasive but often overlooked role, especially in disciplines that are relevant to normative ideas about the proper relations between human sexes, such as biology, primatology, or medicine (among many others).

3.2.1 Biology

A first example of feminist science critiques concerns the **theory of sexual selection**. This is of central relevance, because our understanding of evolution is consequential for many other research fields, such as primatology, evolutionary psychology, or anthropology. Charles Darwin argued that there are two basic mechanisms driving evolution: natural selection and sexual selection. *Natural selection* occurs when variations of inheritable traits in members of a species confer a benefit in terms of chances of survival. Yet some traits, Darwin argued, cannot be explained by natural selection because they are irrelevant or even counterproductive to survival, such as the colourful peacock plumage. Instead, these traits were explained via the mechanism of *sexual selection*: They increase the reproductive success of an animal. Sexual selection, in turn, has two components: first, male–male competition for mating partners; and second, female choice of the most attractive mating partner (Nelson 2020).

Feminist scholars have been critical of Darwinian sexual selection theory from early on (e.g., Bleier 1984; Fausto-Sterling 1985; Hubbard 1983). Pinpointing the problem by asking 'Have only men evolved?', Hubbard (1983) identifies gendered stereotypes in Darwin's work and the implications for questions of gender equality. Darwin presents the male part of the equation as competitive, active, and eager to mate, whereas the females passively watch the male competition and are much pickier about whom to mate with. This, Hubbard argues, is not only androcentric but also in alignment with nineteenth-century Victorian gender roles.

Moreover, Darwin emphasises the role of male–male competition as responsible for the further development of a species, rather than female choice, which he characterises as prioritising beauty without utility (Fehr 2018). This creates unique selection pressures for males, which, ultimately, are thought to result in male superiority. Drawing conclusions on the relation and the differences between the human sexes, Darwin holds that 'Man is more courageous, pugnacious and energetic than woman, and has more inventive genius' (Darwin, *The Origin of Species and the Descent of Man*, cited after Hubbard 1983, 55).

Men are better in many areas because they had to fight harder throughout the history of the species; in particular, this relates to their cognitive capacities.

> The chief distinction in the intellectual powers of the two sexes is shown by man's attaining a higher eminence, in whatever he takes up, than can women – whether requiring deep thought, reason, or imagination, or merely the use of the senses and hands (Hubbard 1983).

Women, it seems, do not contribute much to evolution themselves but merely profit from the fact that men will pass on their superior traits to male as well as female offspring. Darwin's theory of sexual selection has been criticised for being impacted by gender stereotypes, being androcentric, and displaying an insensitivity to the relevance of gender as a social variable regarding sexual behaviour in humans. It thereby perpetuates said stereotypes and contributes to gendered power relations. Its impact also extends into more current biological research. While some biologists such as Ernst Mayr have questioned Darwin's theory of sexual selection and its explanatory import, it also received some support in the twentieth century.

For example, Bateman (1948) argued, based on studies on fruit flies, that variance in reproductive success is far greater in males than in females, motivating males to compete hard for access to reproduction. This was expanded upon in Trivers' (1972) theory of parental investment, which holds that costs associated with the production of (and caring for) offspring differ according to sex. Higher costs for females are then taken to align with, or predict, females being less eager to mate; resulting, again, in a picture of the male as active and promiscuous and the female as passive and coy. Feminists have questioned several central assumptions in these theories. For example, they have inquired into how to operationalise and measure parental investment and associated costs, have warned against overgeneralising empirical results on differential reproductive success rates in specific species, and problematised the identification of sexual activity with reproduction (cf. Fehr 2018; Nelson 2020).[9]

Another famous example is Emily Martin's case study on the gendered language used in research on **human reproduction**. Again, we encounter the stereotypes of the active male and the passive female here, accompanied by a higher valuation of the former. Contemporary textbooks describe the fertile female body as *shedding* an egg each month, with every egg germ cell that never

[9] Sexual selection theory understood in this way, moreover, also informs accounts in sociobiology and evolutionary psychology that consider males and females to be fundamentally different. Some biologists have even taken it to underwrite an understanding of rape as an adaptive reproductive strategy in human males (Thornhill and Palmer 2000; for a thorough discussion and rebuttal, see Travis 2003). Unsurprisingly, sociobiological research has received much scrutiny from feminists (e.g., Fehr 2018; Lloyd 1999; Meynell 2020).

reaches maturity and every unfertilised egg being portrayed as waste, as a failure to reproduce resulting in the destructive process of menstruation. At the same time, the male body is described as *producing* sperm, and scientists marvel at its ability to generate millions of sperm per day, an ability characterised in terms of abundance rather than waste (Martin 1991, 486 ff.).

Gendered metaphors also pervade scientific accounts of egg and sperm cells as such, ascribing stereotypical behaviour to them: The egg passively awaits its penetration and fertilisation by the strongest, fastest competitor among the sperms. At its most extreme, this process is described as a rescue mission for the fragile egg, which would die unless saved by some energetic, brave sperm. Age-old narratives of the damsel in distress to be rescued by the male hero find their way into scientific descriptions here. Human germ cells are unnecessarily gendered, and the perspective chosen is androcentric in that it focuses on the sperm cells' activity, omitting the fact that the unsuccessful sperm will also die (so the process might just as well be described as the egg saving one special sperm) (Martin 1991, 489 ff.).

The problem here is not just that the gendered descriptions of physiological entities and processes perpetuate cultural stereotypes, but also that they lead scientists to interpret observations in a biased way, that is, in a way that leads to systematic misrepresentations and errors. For instance, it was thought that the penetration of the egg was dependent on the sperm's forward thrust, while researchers in Martin's time began to realise that the sperm's force would be mechanically insufficient for this. These researchers began to ascribe a more active role to the egg in the process, which depends, for example, on the presence of specific molecules or microvilli on the egg coat that so far had been overlooked. A more interactive picture arises; however, as Martin points out, this research also sometimes ends up using gendered metaphors, now characterising the egg as a femme fatale catching the sperm, luring it into its trap (Martin 1991, 498 f.). Not only are we thus encountering the phenomenon of *overspecificity* here in terms of describing gender-neutral phenomena in gendered ways, but this is also clearly connected to a double standard in their valuation – the sperm's active role and the magnitude of sperm cell numbers are praised, whereas the egg is either described as passive, or, if active, this activity is portrayed in negative ways.

3.2.2 Primatology

Until the 1970s, research in primatology displayed many of the gendered patterns already mentioned. Primatology is a prime example in feminist philosophy of science because of the paradigmatic shift it has undergone: from being a male-dominated discipline pervaded by androcentrism to becoming what some have called a 'feminist science' (Fedigan 1997). This shift was initiated when more female and/or feminist researchers entered primatology and began to criticise its androcentrism. Up to this point, primatologists had focused almost exclusively on the behaviour of male primates and on certain kinds of behaviour and traits in particular, namely male–male competition, hierarchies structured by male dominance, and male aggression. The familiar picture that resulted – of aggressive, active males and coy, nurturing, passive females – aligned well with, and was informed by, theoretical assumptions from the theory of sexual selection and parental investment theory (Botero 2020).

This androcentric perspective was challenged by researchers who questioned these background assumptions and placed more emphasis on other aspects of primate behaviour that proved equally central to the social structure of primate groups. These included female–female competition, female–male interactions beyond reproduction, female–infant relations, male–infant interactions, female promiscuity, male–male cooperation, and so on. This *widened focus* led to empirical observations that contradicted the account of female primates as passive, unimportant parts of the group's social life (e.g., Fedigan 1992; Hrdy 1981/2009). A further criticism was that earlier primatological research had tended to overgeneralise observations from one specific group (e.g., baboons) to all primates, and from the presumedly general 'primate pattern' to accounts of human nature (Nelson 2020).

The change in focus, moreover, was accompanied by *changes in methodology*. In a germinal paper on observation sampling methods, Altmann (1974) criticised that observational studies of primates often gave no explicit information about their sampling methods. Usually, this meant that they used what she called 'ad libitum sampling', whereby the researcher in the field simply notes as much as possible of what they can observe. As Altmann argues, this may have exploratory, heuristic value, but has little evidentiary import, because it is very prone to bias. All sampling methods require decisions on what to focus on, as observation of primates will not give us a complete account of their behaviour. If such decisions are not made explicit, they often remain unconscious. This can lead to observers recording data on what strikes them as salient – which, in turn, might be informed by background assumptions, gender stereotypes, and so on. An alternative that Altman proposes is 'focal sampling', whereby one observes pre-specified actions in a specific animal or group for a set period of time. This will not necessarily lead to unbiased results, but it introduces a reflexive approach to one's research question, its operationalisation, and observations.

Botero (2020) gives a helpful overview of other methodological changes, including a divergence from the ideal of being a detached, neutral observer and avoiding the threat of anthropomorphism. Instead of observing primate groups from a distanced, hidden place, prominent researchers such as Dian Fossey and Jane Goodall began imitating the animal's behaviour, made themselves visible, and started interacting with them. They also started to refer to the animals by names rather than numbers, all of which was criticised as unscientific. As Botero notes, their new approach mirrored developments in ethnographic methods (such as participant observation) that allow for interactions between researcher and researched and include the researcher's reflection of their own behaviour and its potential impact. Ultimately, the changes in primatology thus extended beyond the abandonment of an androcentric perspective to methods of data collection, critiques of theoretical background assumptions, and a rethinking of the criteria of good science and the ideals of detached, neutral objectivity.

3.2.3 Medicine

As medicine deals directly with human physiology, and as women's supposed inferiority has usually been grounded in biological differences between the sexes, it is unsurprising that both medical research and clinical practice are a rich source for examples of gender bias. As Londa Schiebinger puts it: 'For the most part, academic study of sexual differences was designed to keep women in their place' (Schiebinger 1999, 112). While the concrete explanations of female inferiority via biology have changed over the centuries, the idea itself remains stable. A particularly illustrative example for this pattern is the history of the diagnostic concept of **hysteria**, which has served for millennia to substantiate beliefs about women's general physiological, moral, intellectual, and psychological weakness in comparison to men (cf. also Bueter 2012, 2017).

In ancient Greece, hysteria referred to a variety of possible symptoms (such as headaches, seizures, paralysis, anxiety, and pelvic pain) that were thought to be caused by a wandering uterus, itself triggered, for instance, by childlessness. This aligned with general ideas about the female body as an inferior, incomplete version of the male body due to a relative shortfall in heat, which was thought necessary to attain a higher state of development. In the Christian Middle Ages, this physiological lack was reinterpreted in terms of women's moral and religious inferiority, tracing a line back to Eve's original sin. Symptoms of hysteria became signs of witchery and daemonic possession. In the late nineteenth century, we can see an epidemic of cases of hysteria simultaneously with its resurfacing as a major research topic in medicine. Hysteria was reconceptualised as a disease of the nervous system that primarily affects women but can also sometimes occur in frail, feminine men. This sex distribution was explained by a greater fragility of the female nervous systems, caused by the relation between the nervous system and the reproductive organs. These organs, it was postulated, play a more dominant role in female bodies and lead to a weaker physiological constitution, as well as to a lack of psychological resilience, moral integrity, and willpower (Fischer-Homberger 1979). Accordingly, hysteria was increasingly considered a mental illness, albeit one with biological roots, to be 'treated' by various forms of psychosurgery (such as clitoridectomies, ovariectomies, or hysterectomies). In Freudian psychoanalysis, hysteria became a common problem of neuroticism in women, potentially caused by female penis envy, which blocks the woman's psychological maturation (cf. Tuana 1993, ch. 5). In 1980, hysteria was finally deleted from the official classifications of mental disorders (DSM-III; ICD-10) because of its vague definition and pejorative meaning (for a more complete overview of the history of hysteria, cf. Tasca et al. 2012).[10]

Yet stereotypes of women as less resilient, overly emotional, irrational, or hypochondriac live on, and continue to affect both medical research and practice. Prejudices that women's symptom reports are unreliable and dramatising become visible, for example, when women (as well as people of colour) presenting at emergency departments with chest pain have to wait longer for physician evaluation, are offered fewer diagnostic procedures, and get admitted less often than (white) men (Banco et al. 2022). Symptoms of coronary heart disease also get misdiagnosed as a mental health issue in women twice as often as in men (Maserejian et al. 2009). These are instances of gendered *double standards*.

Issues of **female pain** in general are loaded with biases. Women dominate the populations of chronic pain patients. While the details are not entirely clear, the causes for this seem to be biological as well as psychosocial in kind. Despite sex/gender differences being well-established, studies often do not provide specific analyses when using mixed populations, thereby exhibiting a *gender insensitivity* that makes it impossible to improve our understanding of these differences. One important factor seems to be the relevance of gender stereotypes in clinical encounters. Self-reports of painful experiences are gendered in that there is a 'male' (brave) and a 'female' (emotional) way of dealing with pain, with the former often being considered ideal (Samulowitz et al. 2018). This affects how women get treated in practice: many women report being disbelieved, clinicians describe female patients as hysterical, and women get prescribed less effective treatments (e.g., sedatives instead of analgesics) (Hoffmann and Tarzian 2001).

[10] The explanation for hysterical symptoms and their epidemic rising at certain historical points remains somewhat elusive. The causal background here is probably a very complex one, with multiple psychosocial and biological factors contributing and interacting. For example, Nezhat et al. (2020) argue that many cases of hysteria were actually cases of endometriosis. Endometriosis is a painful and often disabling disease in females that remains underresearched and underdiagnosed, whose sufferers are still often described in gendered ways as 'hysterical', that is, as overreacting (Hudson 2022).

In the medical realm, being female is a double-edged sword that can be connected to both over- and under-treatment. Many feminists have argued that issues connected to female reproduction are **overmedicalised**, that is, treated as a medical pathology even though this pathologizing does more harm than good. Examples are the establishment of Premenstrual Dysphoric Disorder as an official diagnosis in 2013 (cf. Gagné-Julien 2021), the pharmaceutical treatment of low sexual desire in women (cf. Bueter & Jukola 2020), or the conceptualization of menopause as a hormone deficiency disease requiring substitution with synthetic hormones (cf. Bueter 2015). This overmedicalisation can be connected to stereotypes about women being more fragile in general and more affected by their reproductive systems in particular; for instance, the contention that women's hormonal cycles can lead to 'inappropriate' emotional reactions.

Historically, this idea that women are more determined by their reproductive capacities than men has clearly been instrumental in gender oppression. For example, during the first wave of the women's movement in the nineteenth century, when activists were agitating for political equality and the right to education, these stereotypes were used to reject the respective arguments 'on scientific grounds'. The female reproductive system, it was argued, requires so much physiological energy that women cannot simultaneously engage in intellectual pursuits. A woman, in a nutshell, would have to choose between her brain and her womb.

> [A] girl upon whom nature [. . .] imposes so great a physiological task, will not have as much power left for the tasks of the school, as the boy (Clarke 1873/2007, 34).

> [F]emale idiocy is not just [. . .] a physiological fact but also a physiological demand. [. . .] If it were possible to develop the female abilities equally to the male's, the mother organs would atrophy, leaving us with an ugly and useless hybrid. [. . .] Excessive brain activity makes a woman not just wrong but also ill (Möbius 1900, 24.f; my translation).

On the flip side, sex/gender differences in non-reproductive health are often overlooked. As a part of the second wave of the women's movement, a women's health movement (WHM) was formed in the 1960s. The WHM fought for reproductive freedom, criticised the overmedicalisation of aspects of female reproductive health as a means of social control, pointed out paternalism in health care and sexism in medical institutions, and aimed to provide accessible health information for women (e.g., Boston Women's Health Book Collective 1973/1976; Dreifus 1978; Ehrenreich & English 1978). These feminist activists and scholars, moreover, criticised the pervasive *androcentrism* in research on non-reproductive health. At the time, clinical trials were usually conducted on exclusively male populations and the results were generalised and applied to all sexes/genders.

This androcentrism was often accompanied by *andronormativity*. The standard human was conceived of as male, and in consequence, differences in female populations were seen as atypical deviations. A prominent example is coronary heart disease in women. As it became clear that heart attacks in females often differ in their symptoms from heart attacks in males, these were described as 'atypical' symptoms. As a result, heart attacks have been underdiagnosed and undertreated in women, leading to higher mortality rates, and have long been underresearched.[11] In 1993, the NIH Revitalization Act made it a precondition for funding to include women in clinical trials and to perform sex/gender-specific analyses of generated data in the US (setting an example for other countries to follow), in order to avoid androcentrism, overgeneralisation, and gender insensitivity.

While these requirements have been dismissed as an illegitimate politicisation of science by some, recent decades have brought many valuable discoveries about sex/gender differences in areas of non-reproductive health, from oncology to cardiology to immunology. This is relevant to treating the mechanistic physiological basis of diseases as well as their symptoms, in particular with pharmaceutical products. As female bodies, on average, differ from male bodies in terms of, for example, size, body constitution, and metabolism, it has turned out that many medications require different dosages for women, may in general be less effective, or come with a higher rate of adverse effects. Gender roles also have important socio-behavioural impacts on health; for example, when men avoid seeking medical help in order to be tough, or when women allow themselves less time to recover because they have a higher load of care work (for a more detailed discussion, cf. Bueter 2012, 2017; Epstein 2007).

4 Mapping Feminist Philosophies of Science

4.1 Types of Feminist Philosophy of Science

As the previous chapters show, the impact of gender (or, more precisely, of gendered stereotypes and power relations) on science is manifold – reaching from the representation of women in scientific careers to the presence of gender

[11] In 1977, the NIH had excluded women in their child-bearing years from clinical trials in reaction to several tragic scandals of drugs affecting the foetus in utero. In practice, though, this exclusion of females was applied very broadly (e.g., to laboratory animals or post-menopausal women). The guideline was criticised on political grounds for its paternalism as well as an epistemic grounds for its failure to produce generalisable results (Epstein 2007, 303ff.).

bias in numerous scientific theories.[12] Based on these empirical results, feminist philosophers of science have reflected on the implications for our understanding of what science is and should be like.

A minimal answer to the prevalence of gender bias in science would be what Harding (1986, 1992a) calls *spontaneous feminist empiricism* and Kourany (2010) characterises as the *methodological approach* to sexism in science. As a first line of response, this can be found in the work of scientists and science scholars who have been seminal in uncovering gender bias in empirical research (e.g., Ruth Bleier, Anne Fausto-Sterling, Ruth Hubbard, Sarah Hrdy, or Sue Rosser; cf. Kourany 2010, 50–54). Here, cases of gender bias are considered to be cases of bad science, which can be diagnosed as such using current standards of good science. This minimal response sticks to traditional ideals of value-freedom and objectivity while pointing out the presence of gender bias, which needs to be corrected for by stricter adherence to given methodological standards.

Most feminist philosophers have taken more radical stances, however. They argue that there are systematic reasons for the ubiquitous cases of bias and that these call for thorough changes on the normative level, too. They question the traditional ideals of good science as value-free and objective, and replace them by new ideals. What this means in detail differs, and this section, while being far from exhaustive, will introduce some of the most prominent and historically influential approaches. Whereas it remains controversial what the right response to gender bias on the philosophical level is, it is also important to point out that these different feminist philosophies of science share some common ground. In particular, they have a common starting point in an understanding of knowing, and knowers, as (socio-politically) situated, a notion first introduced by Donna Haraway (cf. also Section 2.2).

> Situated knowledges are about communities, not about isolated individuals. The only way to find a larger vision is to be somewhere in particular. The science question in feminism is about objectivity as positioned rationality. Its images are not the products of [. . .] the view from above [. . .] but the joining of partial views and halting voices into a collective subject position that promises a vision of the means of ongoing finite embodiment, of living within limits and contradictions – of views from somewhere (Haraway 1988, 590).

---

[12] It might be objected that gender bias in this latter sense does not pervade all of science; for example, disciplines such as physics, chemistry, or mathematics do not usually encode gendered stereotypes in their theories, as the researched phenomena (e.g., photons, black holes, or mathematical relations) are typically not gendered or perceived to connect to gender relations. Nevertheless, the range of disciplines pervaded by gender bias on the content level (e.g., the life sciences, the social sciences, and the humanities) warrants a rethinking of science in general (cf. also Lloyd 1995b for an application to theoretical philosophy). At the same time, disciplines such as physics, chemistry, or mathematics are among those farthest away from equitable representation and institutional cultures and tend to be portrayed as typically male endeavours (e.g., Schiebinger 1999, ch. 9; Urry 2008).

This contrasts with all tenets of the atomistic model of knowers underlying traditional epistemology and philosophy of science, which conceives of knowers as *individuals* who are *self-sufficient* and *interchangeable* (Code 1991; Grasswick 2004). Drawing on this general insight, one way to map the proliferating field of feminist philosophies of science is in terms of their emphasis on different aspects of this contrast between atomistic and situated knowers (cf. also Grasswick 2018).[13]

A first group of approaches, such as that of Nelson (1990), argue that the proper epistemic subject of scientific knowledge is the *community*, not the individual scientist (cf. Grasswick 2004). A second group of approaches emphasises the *interactive* rather than self-sufficient nature of scientific knowers. These broadly align with what Kourany (2010) characterises as the *social response* to gender bias in science: that is, they build on the insight that science cannot be done by anyone in isolation but needs to be rethought as a communal enterprise. Most feminist empiricist positions fall into this category. A third group of approaches focuses on the *differential* nature of knowers, denying their interchangeability. According to these, not only can no one do science alone, but not everyone can do science just the same way, or have access to all potential perspectives. Different social locations bring with them different experiences and knowledge. Feminist standpoint theories fall within this category. And as the relevant social locations here are placed within a hierarchy of power, standpoint theories are usually not only social but also explicitly political. They fall under what Kourany (2010) characterises as *political approaches*.

Janet Kourany (2008, 2010) herself has also provided a political response to the problem of gender bias in science: the *ideal of socially responsible science*. She starts from the argument that the traditional ideal of value-freedom, too, has had an epistemic and a political role: It is supposed to make objective knowledge possible and thereby to allow for social reform (i.e., by creating unbiased knowledge that can support the feminist call for gender equality). Kourany holds that feminist critiques of the value-free ideal (see Section 5) have shown that the value-free ideal is incapable of fulfilling these roles. And if values in science are inevitable, then science should be guided by social values that

[13] Since feminist philosophy of science has been evolving for several decades now, the field has grown increasingly complex and may be mapped differently. For instance, Harding (1986) has provided an influential tripartite distinction into feminist empiricism, standpoint theory, and postmodernism. I will not go into details of feminist postmodernism here due to its decreasing impact over time (owing to common issues with postmodernism), even though its critical stance on impartiality and rejection of any kind of gender essentialism have been very influential (cf., e.g., Anderson 2020 for more detail). There are also other approaches such as feminist pragmatist accounts (e.g., Clough 2003), which I cannot cover for reasons of space. Kourany (2010) distinguishes between methodological, naturalist, social, and political approaches, which partly overlaps with the distinctions introduced here (see below).

support human flourishing (cf. also 5.3). Scientific success, according to her, needs to be understood in terms of empirical *and* social success.

As this rough taxonomy shows, feminist philosophy of science has developed into a rich and complex research field. In what follows, I will characterise the most influential accounts of the field in some more detail.

4.1.1 Critical Contextual Empiricism and Social Objectivity

Feminist empiricism aims to identify gender bias in the content of science and to improve upon empiricist methodologies to overcome such bias. The empirical reality here plays a role as arbiter and limits what can count as a good theory. At the same time, the role of empirical evidence is considered as complex and often equivocal. Based on a view of theory assessment as holistic, observation as theory-laden, and the thesis of underdetermination of theory by empirical data (see Section 5), *refined feminist empiricists* (e.g., Campbell 1998; Longino 1990, 2002; Nelson 1990; Solomon 2001) focus on how gender stereotypes can affect research and how that might be avoided. In contrast to spontaneous feminist empiricists, they argue that this calls for more than a more rigorous implementation of given standards; rather, it requires a thorough rethinking of these standards, including key concepts such as justification, knowledge, or objectivity.

The most influential position in this field is Helen Longino's (1990) *critical contextual empiricism* (CCE). Critical contextual empiricism does not stop at criticising cases of gender bias in science, but provides a new normative ideal (sometimes called the 'social value management ideal'), central to which is a model of *social objectivity*. Contrary to traditional accounts of objectivity, social objectivity discards the idea of a 'view from nowhere'.[14] This traditional concept of objectivity can be hard to pinpoint, as one can refer to a variety of aspects of science in its name (cf. also Douglas 2004; Lloyd 1995a). For instance, 'objectivity' can be ascribed to (1) *the producers* of scientific knowledge, requiring scientists to be as neutral, disinterested, or detached as possible; (2) *the process* of creating scientific knowledge, via rigorous methods and testing to rule out any biases; (3) *the scientific results* themselves, as correctly representing the world; and (4) *the phenomena* constituting the subjects of said knowledge, as existing in a mind-independent way. In all of these regards, traditional objectivity expresses, and praises, something for being devoid of subjective influences – or of any trace of social situatedness. Accordingly, the value-freedom of science has usually been considered as a necessary condition

[14] For a more detailed discussion of the concept of objectivity, see John (2021).

of objectivity (insofar as values are considered to be subjective in a way that empirical facts are not).

Longino, first, locates objectivity on the procedural level (2). Second, she considers the relevant process to be essentially social. Objectivity, for her, depends on intersubjective criticism from a variety of angles. This critical process can work more or less well, and it can be conducted in a scientific community with more or less homogeneous value-orientations. Objectivity therefore is, third, a matter of degree. It works best in a scientific community that is characterised by a diversity in values and perspectives, as such diversity allows to make hidden background assumptions visible. Thereby, it can help to sift out idiosyncratic value-judgements and biases; yet this does not necessarily result in value-free science.[15] Consequently, social objectivity, unlike traditional objectivity concepts, does not presume (or aim at) value-freedom.

Social objectivity cannot be realised by an individual researcher because it depends on intersubjective criticism. Its essence is not the correct, impartial representation of the world but the *transformative critical process* characterising (good) science. To work well, this critical process requires a certain institutional context. In particular, it depends on the implementation of four conditions related to how scientific communities are to be organised. It requires:

(1) recognised avenues for criticism, such as conferences, journals, and mechanisms of peer review;
(2) shared public standards for criticism, such as epistemic values providing common ground for discussions on theory appraisal;
(3) uptake of criticism, meaning that the community has to be responsive to critical voices;
(4) tempered equality of intellectual authority among diverse participants: that is, the scientific credibility assigned to participants matches their level of qualification rather than correlates with their socio-demographic characteristics (Longino 1990, ch. 4).

This idea of social objectivity has been very influential in feminist philosophy, but has also met with critiques from inside and outside the field. For example, its grounds for rejecting the value-free ideal have been questioned, and there have

[15] Social objectivity is a procedural concept: it locates objectivity in a community's transformative critical process, not in the eventual results of this process. In this, it differs from another account that gives a social response to (gender) bias in science, Miriam Solomon's (2001) *social empiricism*. Solomon distinguishes between, and allows for, empirical and non-empirical decision vectors in theory choice. Among the latter, she includes social factors, such as feminist or sexist values, as 'ideology'. These factors are to be equally distributed between different scientific theories to allow for rational dissent. While not eliminating them, this ideally balances values (or social factors) out.

been calls for the conditions for social objectivity to be more concretely formulated. In particular, the kind and the role of diversity involved have been discussed. One noteworthy line of critique has been whether Longino's account is actually political enough, or even whether it is feminist at all. This is because social objectivity calls for increasing the diversity of values rather than a dominance of feminist values in science (Hicks 2011; Intemann 2010, 2017; see Section 6 for a more detailed discussion).

4.1.2 Feminist Standpoint Theory and Strong Objectivity

Feminist standpoint theory (FST) also stresses the partiality of perspectives and the value of diversity in scientific communities. It builds on the idea that all knowledge is situated: Knowers live in particular socio-historic contexts, face particular material conditions, and occupy specific positions within a society's hierarchy. These circumstances create a variety of social locations that affect both what someone can (or is likely to) know and what their status as a knower is. The distinctive thesis of standpoint theory is that some of these social locations can be developed into standpoints that come with epistemic advantages.

Standpoint theory has its origins in Marxist theory. The idea here is that members of the oppressed classes have epistemic advantages over members of the ruling classes when it comes to recognising and understanding this oppression and its mechanisms. The socially marginalised are thought to have an epistemic privilege that stems from their lived experience. This experience continually confronts them with the reality of power structures, as well as with incongruities between this reality and the dominant ideology.

This general idea has been applied beyond class. For example, Mills (2007) makes a convincing case for an advantage for people of colour in identifying racism in a racist society; after all, one's very survival as a person of colour may require understanding certain things that others may not see. Feminist standpoint theorists inquire into how gender impacts one's lived experience in patriarchal societies, thereby creating perspectives that are socially situated in a particular way and enable an epistemically privileged view of phenomena of gendered oppression. In consequence, here it is not the diversity of perspectives per se that is beneficial to science; it is the presence of socially marginalised standpoints.

Feminist standpoint theory has been controversially debated both inside and outside feminist philosophy of science and has developed substantially in response. It is hard to give a unified account of FST due to its multi-faceted nature. My characterisation here will mainly follow Alison Wylie (2003, 2012), who both gives a helpful overview of historical developments and provides a concise summary of current FST. As she notes, critics of feminist standpoint theory have rejected its supposed essentialism (embedded in the assumption of a 'female' or 'women's' perspective) and have pointed out the intersectionality of the dimensions of oppression and the differentiation of women along the power hierarchy that comes with education, race, economic status, or (dis) ability. In response to such criticisms, Wylie holds that FST does not require, and often explicitly disavows, an essentialist reading of the relevant social categories (cf. also Hundleby 2020, 89). In addition, Wylie does not characterise epistemic advantage as an automatic consequence of inhabiting a certain social location. Rather, it requires one to develop a standpoint, that is, a critical consciousness of one's social location and its relation to other social locations within a power matrix. A standpoint in this sense requires not just a social location but a *political position*.

Being marginalised and oppressed does not automatically come with an understanding of this oppression. For example, in a society structured by gender inequality, people who identify as female might internalise prejudices (such as women being bad at maths) and share a common belief that society is basically just (implying that, e.g., underrepresentation of women in science results not from discrimination but from differences in aptitude). However, daily experiencing the confrontation of such beliefs with the reality of life in a marginalised position makes it more likely that marginalised people will develop a standpoint. This development requires critical reflection and oftentimes a community with shared practices of consciousness-raising, as was common in the women's rights movement.

Such a critical standpoint can confer contingent, local advantages to knowers; that is, the epistemic privilege only holds under certain circumstances and regarding certain phenomena. It is not all-encompassing, but mostly pertains to matters related to social hierarchies. The epistemic privilege of a standpoint thus requires cognitive and emotional work, and it is limited in scope; not least because one dimension of oppression is usually a lack of access to educational and conceptual resources (Wylie 2003, 2012). This thesis of local epistemic advantage can, moreover, be applied to individual knowers or to communities. In the latter case, it holds that a community that includes marginalised standpoints is epistemically advantaged in relevant research contexts (Intemann 2010, 787–89).

Accordingly, social location is not sufficient for a standpoint. Many standpoint theorists also hold that it is not necessary. People in privileged positions are less likely to understand the system they benefit from – but it is not impossible. As Catherine Hundleby (2020, 90) puts it: 'A feminist standpoint is a social orientation politicised through feminism in which any person – not only women – may participate'. Yet respected members of scientific communities are often also relatively privileged in society. To access the epistemic advantage of less privileged people, standpoint theorists provide the methodological maxim of *starting from marginalised lives*. This means that researchers (for instance, in the social sciences) should aim to look at our society 'from below', which will generate critical questions that enable us to understand social processes and structures better. For example, how is employment related to unpaid care and reproductive work? What perspectives are excluded from knowledge practices? Who remains invisible in economic theories? This might be done via interaction with marginalised people, or via careful reflection on the impact of power structures. Doing so, however, is often greatly aided by actually occupying a marginalised position.

This conundrum leads to an emphasis on the role of the *insider–outsider*: people who are marginalised themselves in some respects, but have access to dominant ways of living and thinking at the same time. For example, this might be the poor woman working for a rich family, or the black scientist in a mostly white research environment. Forced to navigate two different worlds, insider–outsiders have immense potential epistemic privilege and can greatly aid research in thinking from the margins (e.g., Hill Collins 1990; Hundleby 2020; Wylie 2003).

Sandra Harding (1991, 1992a, 1992b, 2015) also advocates for a new understanding of objectivity. Feminist standpoint theory's emphasis on the partiality of perspectives and the epistemic privilege of politically mediated, marginalised standpoints leads her to reject traditional 'weak' accounts of objectivity that focus on the impartiality and value-neutrality of scientific contents. Instead, her model of *strong objectivity* not only acknowledges the partiality of perspectives and the value-ladenness of standpoints but turns these into methodological maxims. Rather than aiming for neutrality, scientists, she argues, should start from marginalised lives to uncover and confront hidden background assumptions. A central characteristic of strongly objective science would, accordingly, be its *reflexivity,* which presumes the admission of non-neutrality. It means that scientists need to be aware of, and open about, their ethical and political values and how these impact their research. They should critically investigate their own community and methods, their choice of questions, and their assessment of theories with just as much scrutiny as the phenomena they research. In other words, assuming that we cannot avoid or eliminate value-ladenness, it is reflexivity, not neutrality, that we should aim for. Such reflexivity contributes to a science that explicitly aims for knowledge furthering political goals like empowerment and equality, rather than just knowledge 'for its own sake' that hides its interests behind a veil of detachment.

> [The neutrality ideal] certifies as value-neutral, normal, natural, and therefore not political at all the existing scientific policies and practices through which powerful groups can gain the information and explanations that they need to advance their priorities. [. . .] Thus, when science is already in the service of the mighty, scientific neutrality ensures that 'might makes right' (Harding 1992a, 568 f.).

Strong objectivity is based on the theses that knowledge is differentially socially situated and that epistemic privilege pertains to otherwise non-privileged (but critically reflected) social locations. This can explain both the pervasive existence of gender biases in scientific contents, and the curious fact that research projects with an openly political (i.e., feminist) motivation have been better at uncovering such biases than the presumedly neutral ones. Strong objectivity will thus, according to Harding, lead to scientific results that are less distorted and less false (Harding 1992b).

4.2 Critiques of Feminist Philosophy of Science

From its early years, feminist philosophy of science has often been met with a wholesale rejection of the entire project, usually based on critical arguments that cast it as unnecessary, as threatening the epistemic integrity of science and/or philosophy of science, or as undermining feminist political goals. For one, it has been argued with regard to the case studies mentioned above that while there have been gender biases in previous research, overcoming these required neither more female researchers nor feminism. The idea here is that science rests on certain standards for the conduct of inquiry and theory assessment as well as on critical discourse, which together amount to a **self-corrective process** (e.g., Haack 1993, 37). To consider this self-correction in the case of gender bias as dependent on increasing the input from female or feminist researchers would equal a *genetic fallacy*: Inferring the epistemic quality of a scientific claim from the characteristics of its author (cf. Lloyd 1995b).

In Sections 2 and 3, I showed that it is a central thesis in feminist philosophy of science that institutional, structural issues and content issues are interrelated. While the idea that the contents of scientific theories will be affected by who does the research (and who gets excluded) may seem obvious to some, it appears to reject a central tenet of traditional normative ideals about science. One core idea in science is that it aims for universality in knowledge: that whether something is true or not, or well-founded or not, is independent of who came up with it; and that if it is true, it is true for everyone. Yet the call for more (gender) diversity among researchers is not a call to evaluate claims based on the politics or socio-demographics of individual researchers. It has to be interpreted on a social level instead: If certain perspectives are systematically missing, this can create imbalances in whole fields of research (e.g., by only considering males in medicine or primatology). This may leave us with gaps in our knowledge. Such gaps can become problematic blind spots that mislead our judgements about the evidence (e.g., when we are unaware of the androcentrism in medical research and therefore generalise the results; cf. Bueter 2015).

Uncovering such blind spots helps the advancement of scientific knowledge. Feminist philosophers of science hold that this is furthered by diversity and the inclusion of previously marginalised positions. Critics argue that the relevant changes might have occurred independently of any feminist contributions to the respective sciences, and of course, that is a theoretical possibility. But in practice, the history of science shows that changes to earlier androcentric and sexist frameworks occurred in several disciplines starting roughly at the same time, namely during the second wave of the women's rights movement. To explain this as coincidence seems implausible (cf., e.g., Hrdy 1986). Relatedly, Pinnick (1994) objects to the thesis of potential epistemic privilege conveyed by marginalised standpoints that it is an empirical thesis in lack of supporting data. This seems bewildering, however, if we look at the numerous case studies provided in feminist science studies for gender biases and their overcoming. That empirical work constitutes a solid evidentiary base for significant progress being initiated by including formerly marginalised perspectives.

Other critics have claimed that feminist philosophy of science **undermines feminism** as a political position and is contrary to women's interests. Haack (1996) warns against the creation of a new kind of sexist science in the name of feminism through the perpetuation of stereotypes about feminine cognitive styles. This might discourage women from pursuing careers in the 'hard' sciences symbolically gendered as masculine, or women might feel the need to restrict themselves to feminist (philosophy of) science, effectively marginalising themselves within their discipline. This concern is a valid and important one. However, the hypothesis that women tend to have characteristic cognitive styles, as has been mentioned, has also met with a lot of criticism within feminist philosophy of science and has never been a position widely maintained.

In response to a collection of critical essays on feminist philosophy of science (Pinnick, Koertge, and Almeder 2003), Elizabeth Anderson (2006) has proposed several standards relevant to systematic critiques that address whole fields of inquiry, rather than particular papers or arguments. These are accuracy, perspective, and normative consistency. As uncontroversial as the standard of accuracy seems (i.e., of presenting the current state of the critiqued field correctly and fairly), wholesale rejections of feminist philosophies of science rarely meet it. For example, Anderson details the internal debates relative to Haack's critique of 'feminine cognitive styles' and shows how such claims, made by a few authors in the early times of feminist thinking about science, have long since been discarded.

Koertge (2000) also argues that feminist philosophy of science, as a whole, is a disservice to the feminist cause. According to her, it is self-defeating, because it erodes public trust and the belief in the epistemic trustworthiness of science, that is, in its integrity and objectivity. Such an erosion threatens the rhetorical force of arguments for equality that are based on scientific findings (e.g., on equal intellectual abilities). Koertge's argument is connected to the main concern about feminist philosophy of science: that it aims **to replace truth with ideology**, political correctness, and wishful thinking.

While this concern may apply to some positions in the postmodernist camp that lean heavily towards constructivism, relativism, and a rejection of claims to objectivity, such positions are by no means representative of feminist philosophy of science. Anderson (2006) shows that critiques such as these often target presumed arguments that misread the actual textual basis. For instance, Longino has been attacked by these critics for holding that the thesis of underdetermination implies that we should simply choose between theories according to social values. Yet Longino's position is far more complex. The critique that feminist philosophers undermine science relates to the debate about the proper role of values and political considerations in science, which has also been an issue of germinal concern within the field. I will therefore look at the relevant arguments in more detail in the following chapter. In general, though, it can be maintained that the vast majority of feminist philosophers of science aim to overcome biased science in favour of a better kind of science. Rather than abandoning the ideal of objectivity, for example, they have reworked the concept and thought about ways to augment it.

5 Feminist Accounts of Values in Science

5.1 The Bias Paradox and the Value-Free Ideal

The actual and proper role of socio-political and ethical values in science has been of pivotal interest to feminist philosophers of science. This focus arose from the question of what a replacement of androcentric and sexist science should look like if we were to abandon former normative ideals (such as objectivity understood as impartiality) as mistaken or insufficient. Antony (1993) famously formulated the 'bias paradox': 'Once we've acknowledged the necessity & legitimacy of partiality, *how do we tell good bias from bad bias?*' (Antony 1993, 115; italics in original).

The tension referred to here is the following: feminist philosophers of science want to be able to flag androcentric, sexist bias in science as epistemically problematic, while at the same time doubting the possibility of objectivity as a view from nowhere. If all knowledge is situated and partial, how can research based on feminist values be an advancement over androcentric or sexist research? According to Antony, this tension becomes paradoxical if the ideal of impartial objectivity is being criticised as being in itself partial, in so far as it disguises an androcentric perspective – since the critique of something as partial presumes impartiality as a quality standard.

The bias paradox thus either calls for a new normative ideal replacing impartial objectivity – or for the interpretation of feminist advances in disciplines like primatology as progress towards such impartial objectivity. The majority of feminist philosophers of science have chosen the first option because the ideal of impartial objectivity is not only empirically but also systematically problematic. In the following, I will discuss the problems met by this ideal in more detail.

To do so, I will make a terminological shift and talk about the impact of 'values' rather than 'bias'. Antony equates 'bias' with 'partiality'. Yet 'bias' is often understood as 'systematic error'. But if partiality implies systematic error, it is indeed hard to see how there could be 'good bias'. Values, by contrast, are partial[16] but do not necessarily imply error (Bueter 2022). Talking about values rather than bias therefore avoids predetermining the question at hand – namely, whether, and under what conditions, feminist values can be epistemically beneficial in science.

Feminist critiques of the value-free ideal have inspired an extensive debate on science and values. Here, values are usually characterised very broadly as something considered 'desirable or worthy of pursuit' (Elliott 2022, 3), such as reliable knowledge or gender equality. Biddle (2013) points out that the notion of value is often used in an even broader sense to refer to all kinds of contextual factors. Accordingly, assigning 'value-ladenness' to scientific judgements or decisions means that these are not reducible to either evidence or logic. 'Value' here can refer to a variety of different things, such as goals explicitly pursued by researchers, unconscious stereotypes and prejudices, subjective preferences, or career-related motives, all of which can affect the contents and process of science in various ways.

Science's freedom from such values has traditionally been considered as a necessary condition of objectivity. This does not require science to be completely detached from any kind of value consideration but only that the *epistemically relevant*, *internal parts* of science are value-free. This difference between internal and external aspects of science has often been spelt out in terms of the context distinction. At the heart of science lies the *context of justification*, where we test a hypothesis, theory, or model, and assess its validity. This assessment is

[16] I consider values as partial in the sense that members of scientific communities are likely to differ in their actual value-orientations. It might be that some values, or value-judgements, are objectively right; cf. Section 6.2 for further discussion.

supposed to be epistemically independent from decisions traditionally relegated to the *context of discovery*, such as the choice of research questions and the initial formulation of hypotheses. Here, the value-free ideal allows for values that affect the direction of research. According to this line of thinking, one may posit a sexist hypothesis on cognitive differences between men and women, but the assessment of whether the empirical data support that hypothesis needs to be value-free. The results may, moreover, lead to applications in the real world. Deciding whether and how to implement and communicate certain results in the *context of application* is, again, value-laden.

Regarding the epistemic core of science, there is an additional complexity. With the demise of both logical empiricism and falsificationism in the latter half of the twentieth century, it became more and more obvious that the justification of theories, as well as the choice between different competing theories, is not algorithmic. Evaluating the validity of a theory cannot be reduced to matters of logic and data but calls for additional criteria such as simplicity or explanatory scope. These criteria have been called cognitive or epistemic values and treated as legitimate in the context of justification – they are, so to say, scientific values, as opposed to non-epistemic (e.g., political, ethical, or religious) ones.

The current value-free ideal (VFI) thus holds that non-epistemic factors legitimately impact decisions in the contexts of discovery or application, and can affect methods for ethical reasons (in particular, regarding experimentation with living subjects). In the context of justification, however, where we draw inferences from empirical data and assess a theory's epistemic merits, only epistemic values are allowed. As Heather Douglas neatly summarises it, 'the value-free ideal is more accurately the "internal scientific values only when performing scientific reasoning" ideal' (Douglas 2009, 45).

As described above, empirically oriented science studies have demonstrated with numerous examples that science does often not live up to the VFI. Prima facie, this need not to affect value-freedom as a normative ideal, however – so long, that is, as pursuing this ideal and approximating it is considered possible and desirable.

As Elliott (2022) points out, there are three main reasons given in the literature for why one should pursue value-freedom in this sense. First and foremost is the worry that allowing for non-epistemic values in theory assessment will lead us astray in our search for the *truth*. For example, this is the primary concern of Koertge's (2000) defence of the VFI, in which she argues that we need to keep scientific goals (such as explanation) and political goals (such as emancipation) distinct in order for the latter not to undermine the former. A second worry stems from the role that science plays in democratic societies that incorporate scientific results into their political decision-making. Such decision-making arguably requires a neutral, value-free evidence base in order to be democratic. If the evidence base itself is shaped by values of the scientists involved, these will implicitly taint political decisions without any democratic legitimization of the values in question. Thereby, the *autonomy* of the decision-making based on such information will be threatened, be this on the public or individual level (Betz 2013). The third motivation for the VFI is related to this and points to the role of *public trust* in science. If scientific research is not neutral, and evidence-based policy is not autonomous, the public might lose trust in science – and might even be normatively warranted in their distrust (cf. Bright 2018).[17]

Truth, autonomy, and trustworthiness in themselves are things that we value and would not give up lightly. However, many feminist philosophers have argued that there are systematic problems with the VFI that are so pervasive as to call for normative alternatives. In the following, I will present some of these arguments: the so-called gap argument, based on the thesis of underdetermination; the rejection of the distinction between epistemic and non-epistemic values; and the argument from blind spots.[18]

5.2 Underdetermination and the Gap Argument

The current VFI, understood as 'epistemic values only in scientific reasoning', presumes a hypothetico-deductive (HD) model of scientific testing. In such a model, scientists propose a hypothesis, from which empirical consequences are deduced (i.e., predictions about what the data ought to look like if the hypothesis were true). Scientists then gather empirical data from observations or experiments that allow for inferences about the status of the hypothesis by lending it inductive support or by contradicting it. This oversimplified image of the scientific method has been called into question both by attending more carefully to actual scientific practice and by systematic arguments such as the underdetermination thesis, which has played a pivotal role in feminist rejections of the VFI by grounding the so-called *gap argument*.[19]

> This gap [is] between what is present to us and the processes that we suppose produce the world that is present to us, between our data and the theories, models, and hypotheses developed to explain the data (Longino 2002, 125).

[17] Holman and Wilholt (2022) similarly argue that the value-free ideal is supposed to promote the goals of veracity (the discovery of knowledge), universality (ensuring usability by everyone, independent of their value-orientation), and authority (by warranting public trust in science).

[18] This is not an exhaustive list. For reasons of space, I will not discuss the argument from inductive risk (Douglas 2000) nor from an irreducible value-ladenness of certain concepts (such as 'rape' or 'well-being') here.

[19] Cf., for example, Anderson (1995), Kourany (2003), Longino (1990, 2002), and Nelson (1990, 2022). To keep it (somewhat) brief, I will concentrate mostly on Longino's treatment of underdetermination here.

The underdetermination thesis has been discussed very controversially in the philosophy of science, not least because it can be interpreted in different ways and with different degrees of strength. It is helpful to start with a distinction between contrastive and holistic underdetermination here. *Contrastive underdetermination* is the idea that even if we have a theory that is empirically adequate, there could always be alternative theories that fit the empirical evidence equally well, yet are incompatible on the theoretical level; that is, they make conflicting claims about the unobservable realm (Quine 1975). This version of the underdetermination thesis has played an important role in debates on scientific realism.

In feminist philosophy of science, *holistic underdetermination* has been central. Holism here refers to the complexity often involved in scientific testing due to the epistemic embeddedness of hypotheses in a theoretical and methodological framework. This casts doubt on the idea of a direct comparison of hypotheses with the state of the world via the relevant empirical data. To derive empirical consequences from hypotheses, we need to make use of a wide variety of auxiliary assumptions (Duhem 1906/1954). Holistic underdetermination points to the fact that even if there is a conflict between predictions and data, this might be due, for example, to a mistake in the theoretical background used to derive the said prediction, to a measurement error, to inadequate study design or implementation, or to a misleading interpretation of the data. In response to recalcitrant observations, one might thus make changes at different points of this holistic net to ultimately achieve empirically adequate theories. Quine (1953) has famously argued that those changes could in principle affect all beliefs or propositions relevant to the holistic net, including those that used to be considered analytic and necessary truths.

The question that arises is whether all such changes are equally legitimate, and on what grounds we can evaluate the epistemic merits of different alternatives. The problem of underdetermination brings with it the spectre of irrationality. Very strong interpretations of the underdetermination thesis argue that if empirical data cannot solve situations of underdetermination, contextual factors or social values will, instead – and therefore, science would reduce to politics by other means. Feminist philosophers of science usually rely on more moderate interpretations of the consequences of underdetermination. Prominently, Longino's critical contextual empiricism is based on an underdetermination thesis concerning evidential relations. Longino argues that there is a distance between hypothesis and data because we need to invoke auxiliary assumptions (or, in her terminology, background assumptions) to decide what data will be most relevant to the evaluation of a hypothesis, as well as to interpret the data and assess its implications (cf. Longino & Doell 1983; Longino 1990, ch. 3.; Longino 2002, ch. 6). A corollary of this is that relevant situations of underdetermination do not necessarily require empirically equivalent alternatives, but also concern situations where we have alternatives with empirical strengths in different areas, that is, situations of conflicting assessments of what the most significant empirical consequences are.

According to Longino, the gap between data and theory is bridged by background assumptions. Such background assumptions can involve theoretical convictions, methodological principles, metaphysical ideas – and, potentially, social values, stereotypical beliefs, or individual bias. Not all background beliefs will be equally legitimate, as it is possible to give justifying reasons for their employment in specific situations. This justification, however, is not simply dictated by empirical data, either. Moreover, the assumptions used in closing the gap between theory and data are often not made explicit but taken for granted, frequently remaining unconscious. Dealing with the problem of underdetermination requires to make these assumptions visible and thereby debatable. For Longino, this is best done by a scientific community that fulfils the conditions of *social objectivity* (see Section 4) and brings a variety of background assumptions to the table (Longino 1990, ch. 4).

A famous example stems from evolutionary anthropology, which studies the anatomical and social evolution of humans (cf. Longino & Doell 1983; Longino 1990, 105–111). Relevant data are sometimes hard to come by in this field (e.g., fossils of skeletons, teeth, footprints, or artefacts) and usually require interpretation. The evidential relevance and theoretical implications of the given data are therefore often contested. Other sources of evidence are even more controversial, such as data about the behaviour of humans in contemporary hunter-gatherer communities, or the behaviour of non-human primates, both of which are sometimes used to draw inferences about the behaviour of early humans.

A pivotal question has been the development of tool-use in humans. Traditionally, this was answered by the 'male-the-hunter' theory, which attributed the development of tools to their use in big-game hunting by males. This picture is reminiscent of other gendered accounts in biology and primatology insofar as it assumes that it is a stereotypically male activity that becomes the driver of evolution. Starting in the 1970s, this account has been challenged by the alternative of a gynocentric 'women-the-gatherer' theory, which attributes the development of tool-use to women. The proponents of this account focused on other kind of artefacts (e.g., wooden versus stone tools) and argued that tools were invented to deal with the greater nutritional stress caused by pregnancies and the nurturing of offspring. In this picture, women became the active party and made technological inventions related to stereotypical female activities such as gathering, preparing, and storing food.

The point here, for Longino, is not whether the gynocentric account is better than the androcentric one. Rather, the availability of two conflicting accounts shines a spotlight on the background assumptions at play in the determination of evidential relevance and the explanation of data. This also makes the gendered nature of these background assumptions visible, something that prior to the development of a competitor to the male-the-hunter interpretation remained unacknowledged. The competitors here also share certain value-laden background assumptions, namely that there has been a gendered division of labour that matches current ideas about typically masculine versus feminine activities. More recent research calls this into question; for example, by discoveries of female skeletons that have been buried with hunting artefacts (e.g., Haas et al. 2020).

Note, though, that this is not a situation in which an underdetermined theory choice (between man-the-hunter and women-the-gatherer) has now been resolved by further incoming data. It is sometimes argued that situations of underdetermination are often transient, in the sense that further research will discover the decisive evidence. In turn, this would mean that the appropriate response would not be to give up the value-free ideal by allowing for value-laden background assumptions, but rather to remain agnostic until the relevant evidence is in (e.g., Kitcher 2001, ch. 3; Haack 1993). Yet all new data will be equally affected by the gap between theory and data. For instance, the new findings of female skeletons need interpretation, too. Are those artefacts actually hunting tools? Were women buried with hunting tools because they themselves were hunters, or were they maybe wives of important male hunters? And even if there were female hunters, maybe they were rare exceptions of little evidential importance to the general man-the-hunter thesis?

For Longino, in consequence, the underdetermination of theory by data undermines the value-freedom of science by creating an entrance for value-laden background assumptions that can affect the interpretation of empirical findings and assessments of their relevance (and thereby the justification of theories). A more adequate ideal for science would then be *social objectivity,* the implementation of a fair and transformative critical process in a diverse community. This would help to identify value-laden background assumptions and to screen them out where they are deemed problematic. However, this cannot guarantee value-freedom, as value-laden assumptions shared throughout the scientific community will persist: 'And while the marks of individuals may be eliminated by this process, the marks of the culture are not' (Longino 1990, 224).

It seems, then, that Longino rejects the value-free ideal because, first, the gap between data and theory is often filled by value-laden background assumptions, and second, even an ideal transformative critical process cannot be thought of as ultimately closing the gap. Rather than aiming for value-freedom, social objectivity utilises the values present in the scientific community, because it is the clash of different values that will bring them into the limelight. Yet holding that values can have a positive role on the community level insofar as they help identify and eliminate more idiosyncratic values seems to concede the desirability of value-freedom. So why not take value-freedom as an ideal that is worthy of pursuit and that science should try to approximate even if complete value-freedom will remain elusive?

In response, Longino proposes that the values ideally remaining and impacting scientific contents would be vetted in a critical process, too. They will be shared not just as a consequence of their invisibility but as an outcome of the discussion (Longino 1990, 216). This raises the question, however, of what constitutes the grounds for considering surviving values as legitimate, and whether that reduces to community agreement on these values.

A related concern is formulated by Intemann (2005), who argues that the argument from underdetermination is unable to do what feminist philosophers of science want it to. The existence of a gap between theory and data as such, she argues, does not imply a particular normative way of dealing with it. Intemann concedes that this gap exists and is a potential entrance for value-laden assumptions. But what she argues is lacking is an argument to the effect that such value-judgements can be *good reasons* for making a particular theory choice.

5.3 Epistemic versus Non-Epistemic Values

The gap argument against the value-free ideal gains in strength if coupled with a rejection of the distinction between epistemic and non-epistemic values. One plausible response to the problems created by the underdetermination of theory by data is to separate empirical and evidential equivalence: Theories can be equally compatible with the empirical data, but this does not imply that they are equally good theories. Thus, what is needed are criteria to choose between empirically equivalent theories. If these criteria can be considered purely scientific or epistemic, the VFI might be saved in the sense of at least being pursuit-worthy. If, on the other hand, such criteria of theory choice are not independent of social values and contextual factors – and yet can be epistemically legitimate – then defending the VFI even as an aim we can at least gradually approach becomes problematic.

Kuhn (1977) proposed a by now iconic list of criteria to guide theory choice: empirical accuracy, internal consistency, external coherence with other accepted theories, explanatory scope, simplicity, and fruitfulness. According to him, the history of science shows that scientists often invoke these characteristics when deciding between theories (or, in his case, paradigms). At least to some extent, then, the choice between paradigms can rely on paradigm-independent criteria;

which, in Kuhn's account, enables such choices to be rational. However, these criteria are not hierarchically ordered, may conflict with each other, and require interpretation when applied to specific theory choice situations. Is empirical accuracy more important than simplicity? Does simplicity mean mathematical ease or ontological frugality? Questions like these show that theory choice is not algorithmic, even given a shared list of criteria. This is why Kuhn speaks of *values*: because they need to be interpreted and weighted, they can guide, but not determine theory choice.

McMullin (1983) has coined the term 'epistemic values' for Kuhn's list, and distinguishes them from non-epistemic ones.[20] The characteristic underlying this distinction is that epistemic values are supposed to promote the truth-like character of science (McMullin 1983, 18): Only theories displaying epistemic values will survive rigorous scientific examination and stand the test of time. Like Kuhn, McMullin does not provide systematic arguments for the values that made his list but takes them to be implicit in scientific practice. 'The characteristic values guiding theory choice are firmly rooted in the complex learning process which is the history of science; this is their primary justification, and it is an adequate one' (McMullin 1983, 21).

This justification, however, presumes that the history of science is (at least in large parts) a history of moving towards truth-likeness, or of being a general success story. As a feminist science scholar, one will tend to be somewhat sceptical about this presumption, given the myriad of case studies of theories displaying androcentric, sexist, ethnocentric, or racist bias in modern-day science. Given that these theories were chosen for being characterised by epistemic values, the history of science rather seems to show that epistemic values do not provide protection against bias.

In fact, feminists have questioned the very distinction between epistemic and non-epistemic values and the assumed role of that distinction in keeping 'non-scientific' values out of science. To begin with, Rooney (1992; cf. also 2017) pointed out that one can find varying lists of epistemic values in the history and philosophy of science, indicating a lack of clarity in the distinction. Moreover, as already pointed out by Kuhn, epistemic values need to be interpreted and weighted. Their application in specific theory choice situations, therefore, is affected by background assumptions. These background assumptions can be value-laden, which might lead us to rate and read a given list of epistemic values in a way that will turn out to prefer the androcentric alternative. In this way, androcentric values (or interpretations of epistemic values based on androcentric

[20] This distinction is also often designated as 'cognitive' versus 'non-cognitive' values.

background beliefs) can become integral to our understanding of what it means for a theory to be a 'good'.

Distinguishing epistemic and non-epistemic values, in consequence, will not make science value-free, even if we rely on epistemic values only in theory choice. Longino (1996) builds on Rooney's critique and argues against the very possibility of a context-independent distinction between epistemic and non-epistemic values. Her argument can be reconstructed as depending on the following premises:

(1) A value is epistemic if and only if theories that display this value are more likely to contribute to the goals of science.
(2) The goals of science are various, since science aims for significant truths.[21]
(3) Significance can have social/practical and epistemic aspects.

Given these premises, she argues that scientific goals can have a mixed social/epistemic character. For example, feminist scientists are pursuing such mixed goals when they are interested in understanding mechanisms of gender oppression, or in uncovering androcentric assumptions in earlier theories of their field. This holds not only for feminist research: the goals of science involve significant ascriptions that often depend on social as well as epistemic concerns. Therefore:

(4) The goals of science are of a mixed social/epistemic character.

Epistemic values can then be determined via their contribution to such mixed social/epistemic goals. What criteria enter our list of epistemic values thus depends on our respective goals – and this leads to the possibility of variable lists of epistemic values. Longino (1996) proposes the following list of values as contributing to feminist epistemic goals: empirical adequacy, internal consistency, novelty, ontological heterogeneity, complexity of relation, applicability to human needs, and diffusion of power. She contrasts this list with Kuhn's to demonstrate how such values can contribute to socio-political and/or epistemic goals depending on the context. For example, Kuhn lists the coherence with

[21] I am using Philip Kitcher's terminology here. Kitcher (2001) has argued that even if we presume that science aims for truth, this is insufficient to direct inquiry: we are not interested in a collection of any truths whatsoever, but in truths that help us understand, explain, predict, or manipulate parts of the world in ways that are meaningful to us – in short, *significant* truths. Such significance can have practical reasons and/or be grounded in social values. For instance, we might be interested in truths about the human immune system's reaction to a certain virus, because we want to cure or control the disease this virus causes. Longino (2002) also argues that knowledge as a goal of science underdetermines research direction as we seek particular kinds of knowledge about particular things. Longino (2008, 80) likewise argues that '[t]ruth simpliciter cannot be such a goal, since it is not sufficient to direct enquiry. Rather, communities seek particular kinds of truths'.

other accepted theories in a field as an epistemic value; by contrast, novelty favours those theories that do not fit with the current consensus. Arguably, novelty is indicative of success if one aims to overcome pervasive androcentric or sexist bias in a field, whereas external coherence in the same situations will perpetuate such biases. Therefore:

(5) What is an epistemic value varies depending on the goals pursued and on the sociohistorical context of the research.

The point of Longino's argument is not that the Kuhnian values should, in general, be replaced by her list of feminist epistemic values. Neither does she argue that in cases of theory choice characterised by underdetermination, we should adopt the more politically progressive theory (cf. Kourany 2003 for such an argument). Instead, the upshot is that a value can serve epistemic and/or social functions depending on the particular goals and context. Thus:

(6) There is no absolute, context-independent distinction between epistemic and non-epistemic values.

In accordance with this, many of the case studies presented in Section 3 can be interpreted as exhibiting changes not only on the levels of data and theory but also of goals, resulting in new standards for good theories. For example, women's health research pursues a goal that is epistemic (understanding women's health better) and political (improving gender equality) at the same time.[22] This goal has been furthered by research that aims for ontological heterogeneity rather than (an ontological interpretation of) simplicity. Simplicity had been used to support the exclusion of women from clinical trials (e.g., to avoid hormonal changes complicating the results); ontological heterogeneity, by contrast, discourages the underlying idea of a standard human and thereby helps overcoming androcentrism that is problematic, epistemically and politically. Advances in primatology have also been accompanied by changes on the methodological level and a rejection of formerly dominant epistemic values, such as the avoidance of anthropomorphism. As Botero (2020) explains, anti-anthropomorphism had traditionally been considered a core value to ensure that the interpretation of animal behaviour is scientific and unbiased by human preconceptions. Animal behaviour was supposed to be interpreted in the way that assigns it less cognitive complexity, and anti-anthropomorphism was defended in terms of simplicity: ascribing less cognitive complexity was

[22] One might argue that these goals are not on a par, as the latter were not a properly *scientific* goal. The point is not that political goals should replace epistemic ones but that being epistemic does not make something apolitical, since the choice of particular epistemic goals is intertwined with considerations of social significance.

assumed to be better in epistemic hindsight. Feminist primatologists, however, rejected this value in favour of a more complete understanding of primate behaviour, and of overcoming earlier blind spots in primatology. Both the rejection and the allowing of anthropomorphism here are argued to contribute to epistemic goals; at the same time, both are also connected to value-laden assumptions about human exceptionality.

To sum up, Longino argues that what makes a value epistemic is its contribution to scientific goals. As such goals are variable and often simultaneously social and epistemic, resulting lists of epistemic values will also vary. Restricting values to epistemic ones, therefore, will not necessarily result in freedom from non-epistemic values, since what counts as an epistemic value depends on value-laden decisions about the appropriate, or most important, goals of research in a given field. As long as we consider it legitimate to pursue particular epistemic goals for social reasons, we cannot retain the ideal of value-freedom by restricting legitimacy to purely epistemic values neutrally applied, as there is no such thing.

A problem resulting from this is a potential lack of shared standards for theory assessment (which are an integral part of Longino's own conception of social objectivity). If the standards of theory assessment depend on particular research goals, researchers pursuing different goals may end up without a shared basis for discussion. Whereas Longino explicitly allows for a pluralistic landscape of local epistemologies (Longino 2002, chapter 8), it seems that a transformative critical process needs to be possible beyond localised research communities. In particular, this follows from social objectivity's dependence on critique from various perspectives and on the grounds of various background assumptions (Bueter 2010). Social objectivity, then, demands a joint striving for consensus. This, however, does not require a fixed list of epistemic values for all of science. Social objectivity can allow for variety in such lists as long as these are also subject to a process of transformative criticism that draws on a range of perspectives. As Carrier (2013) argues, varying lists of epistemic values can be accommodated if the overall scientific community shares an 'epistemic attitude' that ensures a striving for consensus and enables rational discussion. This epistemic attitude is not operationalised in terms of standards for theory appraisal, but rather in terms of procedural standards of how to debate knowledge claims.

5.4 Scientific Ignorance and the Argument from Blind Spots

Even if we grant the distinction between epistemic and non-epistemic values to proponents of the VFI, using only epistemic values in theory choice might be insufficient to sift out non-epistemic values and bias as it is intended to do.

This is so because the choice itself might be skewed. Okruhlik (1994) argues that insofar as theory choice is comparative and we choose the best of a given number of alternatives, epistemic values won't help against bias: The best candidate (according to epistemic values) of a pool of sexist theories is still a sexist theory. This problem could only be avoided if no theory that is empirically adequate and exhibiting epistemic values to some degree could be biased. Yet this defence seems elusive in light of the many case studies on biased theories previously deemed acceptable and defended via recourse to epistemic values by contemporary scientists.

While Okruhlik analyzes this problem in terms of a choice between existing theories, it has been argued that the problem goes beyond that. Elliott and McKaughan (2009) draw attention to how the generation of data too is affected by the hypotheses and theories one wishes to test. In fact, even the value-free ideal allows for non-epistemic values in the contexts of discovery and pursuit; that is, they influence decisions on what research projects to work on and thereby on what data to gather. Yet the evidential relevance of data is often not uniquely determined by a theory but mediated by background assumptions about, for instance, the salience of variables. Consequently, a theory might be well supported by the data at hand – until a rival theory leads to the generation of additional, conflicting data. To illustrate this, Elliott and McKaughan draw on a case study on the toxicity of certain substances. As long as one presumes there is a threshold for toxicity, data on lower levels of exposure might simply not be considered and all the generated evidence might support the model, until research based on different models leads to additional data showing effects below the assumed threshold.

One might counter this issue by adding a requirement of completeness of evidence to the list of epistemic values. Of course, deliberately ignoring potential data that might undermine one's claims is at odds with any claim to epistemic integrity. However, assessing whether all relevant data has been covered can be tricky, especially when the theoretical alternatives that would make additional data seem relevant are not given. In Bueter (2015), I have called this the argument from blind spots. Essentially, it points to the relevance of unconceived alternatives (Stanford 2006). The assessment of a theory's evidential merits is dependent on the data considered relevant and on comparison to competing theories. Both of these factors might look very different if we had a vaster array of theories and data to consider – just as the development of feminist alternatives served to make inadequacies in earlier androcentric theories visible. Theory assessment, then, is affected by unconceived alternatives.

Assuming the plausible premise that science should look not for each and every truth but for significant truths – and that, in practice, it needs to do so as long as its resources are limited – we will always face a need for selectivity in agenda setting and will consequently ignore other potential lines of inquiry. From the perspective of feminist philosophy of science, this becomes problematic when such ignorance is not randomly distributed but systematically excludes research inspired by feminist rather than sexist or racist values, and when it addresses the interests of the wealthy and powerful rather than the needs of marginalised and oppressed groups.

The argument from blind spots has two important corollaries. First, the current VFI allows for non-epistemic values in agenda setting, but not in theory assessment. It thus presumes that theory assessment is epistemologically independent of decisions related to the choice of research projects and questions. However, given the relevance of unconceived alternatives to theory assessment, this central presumption does not hold, and this in turn calls the conceptual soundness of the VFI into question. Second, this puts agenda-setting and theory pursuit more firmly on the philosophical research agenda. One advantage of a more diverse community is that it will likely produce a wider variety of hypotheses to test, and display a wider range of research interests.[23] In line with this, Harding's model of *strong objectivity* explicitly takes the context of discovery into account. Rooted in standpoint theory, it includes the methodological imperative to start from marginalised lives, that is, to target problems that affect people belonging to marginalised groups and to take their perspectives as starting points. This could help to identify and overcome former blind spots in research.

This also implies the relevance of scientific ignorance as a research subject, one that so far has not received much attention in philosophy of science. Central questions in a *philosophy of scientific ignorance* include how to figure out where our systematic blind spots lie, what causes them, and how to overcome them. It connects to feminist concerns about our lack of knowledge about phenomena that mostly affect marginalised groups and, in general, about the distribution of research efforts and results (cf. Fernández Pinto 2020 for a helpful overview). What kind of things don't we know about? What do we fail even to investigate? What knowledge might get lost, suppressed, or belittled? In her pioneering work on the epistemology of ignorance, Nancy Tuana (2004, 2006) has provided a taxonomy of different kinds of ignorance and illustrated how they relate to feminist issues such as knowledge about women's health or female sexuality (or, we might add, any kind of sexuality that is not of the presumably standard, male kind). For example, she distinguishes between things we don't know

[23] Arguably, such a diversity can also be lacking in thoroughly commercialised fields (such as nutrition science), which negatively affects the epistemic trustworthiness of results (Jukola 2021).

because we are simply not interested in them, things we don't know that we don't know, things we actively ignore (such as, e.g., our own privilege in a sexist or racist society), or knowledge that is systematically undermined in order to maintain or foster people's ignorance. The latter phenomenon might, for example, be found in the suppression of evidence on the harmful effects of medications or other substances due to commercial interests, or in the deliberate manufacturing of doubt regarding scientific knowledge about anthropogenic climate change (cf. also Biddle and Leuschner 2015 on epistemically detrimental forms of dissent). As this indicates, dealing with scientific ignorance requires us to distinguish different kinds, sources, and remedies, as well as to discuss which cases of ignorance we should overcome and which we should accept.[24]

5.5 Beyond Value-Freedom

At the beginning of this section, I pointed out the three motivations for the value-free ideal given by Elliott (2022): to ensure science stays on the path to discover truth; to enable autonomous, democratic decision-making based on untainted information; and to thereby enable public trust in science and science-based policy. Given the argument from underdetermination, it is not clear that the VFI's claim to promote truth can be upheld, as even a strict adherence to its standards may leave us with biased interpretations of data themselves and of their evidential impact. Invoking the distinction between epistemic values and non-epistemic values is supposed to solve this issue, but is argued to be itself problematic, as the application of such values might be affected by non-epistemic ones and the very distinction might depend on value-laden research goals. Moreover, the VFI's focus on the justification of theories is untenable in light of the argument from blind spots. Aiming for value-freedom, then, might not get us any closer to the truth. At the same time, none of the alternatives formulated in the feminist literature proposes simply replacing facts with values; rather, they target how we can deal with the impacts of non-epistemic values in a responsible manner that preserves epistemic integrity.

Consequently, it is also not clear why the VFI should be better than these alternatives at enabling autonomy and trustworthiness. In particular, the VFI is unable to handle the problem of blind spots and their epistemic consequences. Critical contextual empiricism or standpoint theory, by contrast, could address this by increasing diversity and representation, or by focusing on marginalised perspectives and addressing the needs of the powerless. Rather than understanding autonomy and trustworthiness as universal values, it might be fruitful here

[24] See also Proctor and Schiebinger (2008), who coined the term 'agnotology' for the study of ignorance.

to ask whose autonomy is served, and whose trust is warranted, by a science that is only presumably value-free, versus one that explicitly integrates egalitarian values. The worry that a science explicitly devoted to feminist, non-racist, non-ableist, or non-classist values will be distrusted by certain groups is well-founded. Yet if neutral truth is not on the table, science that is shaped by sexist, racist, ableist, or classist values might earn the trust of those dominant groups while failing other people. To circle back to Antony's bias paradox, the important question may then be not how to distinguish good from bad bias, but how to best deal with the unavoidable bias (or value influences) that remain. Answers to this question often focus on the utilisation of diversity, which will now be discussed in more detail.

6 Diversity and Epistemic Justice

While feminist philosophers of science agree on the epistemic benefits of diversity in science, there is considerable dissent on the kinds of diversity required, as well as on its limits and implementation. Intemann (2010) points out that the two most influential approaches in the field – feminist empiricism and standpoint theory – have incorporated many shared insights and converged towards compatibility. Both provide social, contextualist, and normative epistemologies that integrate feminist aims, reject the value-free ideal, and emphasise the value of diversity in scientific communities. Intemann argues that with regard to the disagreements that remain, feminist empiricists should adopt the standpoint theoretical positions, which could create a consensus on *feminist standpoint empiricism.* These disagreements concern (1) the kind of diversity needed, and (2) the role assigned to feminist values. In this section, I will discuss Intemann's proposal, identify some remaining problems for reconciling feminist empiricism and standpoint theory, and argue that a consensus might be reached via a focus on implementing diversity in an epistemically just manner.

6.1 The Kinds of Diversity

The first disagreement Intemann (2010) identifies lies in the kinds of diversity prescribed by feminist empiricists (particularly, Longino's critical contextual empiricism) versus standpoint theorists. Whereas Longino primarily calls for a diversity of values, standpoint theorists focus on diversity of situatedness, that is, diversity of the social background of the members of scientific communities. Both, moreover, agree that social diversity in science is an important political goal, and that such diversity is epistemically beneficial. Yet they justify the latter claim differently.

In CCE, background assumptions play a critical role due to the underdetermination of evidential relations. This is an entry-point for value-judgements that might lead to problematic, biased results, if they go unnoticed. To make background assumptions visible, criticism from a variety of viewpoints and value-orientations is helpful. Shared value-laden background assumptions are often taken for granted and remain implicit. To identify such value-judgements is easier for those who don't share them, as the difference in values will create friction between one's background beliefs and the contemporary consensus in the scientific community.

Feminist standpoint theory calls for a diversity in terms of social locations rather than values. It starts from the thesis that all knowledge is situated, that is, embodied by knowers with different experiences based on their living conditions in a society structured by power relations. Intemann (2010, 2017) argues that it is this difference in experiences that enables criticism of shared background assumptions and provides epistemic benefits to the community. Moreover, FST does not just call for any kind of diverse scientific community, but ascribes epistemic advantages to communities that integrate researchers from marginalised groups.

> The claim is that members of marginalised groups are more likely to have had experiences that are particularly epistemically salient for identifying and evaluating assumptions that have been systematically obscured or made less visible as the result of power dynamics (Intemann 2010, 791).

The first question, then, is whether epistemically beneficial diversity requires a diversity of values or of social background. An easy answer that can resolve the conflict comes to mind: why not both? In fact, kinds of epistemically beneficial diversity can also be argued to go beyond these two categories to involve differences in research strategies, skills, interests, theoretical orientation, and so on (cf. Rolin 2017). If the goal is to identify implicit value-laden background assumptions to avoid bias, then different kinds of factors might all be relevant: friction with one's values or one's experiences can both be helpful here. For instance, one might come to doubt capitalist background assumptions relevant to sociobiological theories that cast economic inequalities as 'natural' based on one's political commitments to economic equality. One might also come to doubt the same assumptions based on one's lived experience of poverty which is extremely hard to escape no matter one's level of effort or ability. It seems unwise to exclude either of these routes if they can both lead to the result aimed for.

An underlying question here is how these different factors relate to each other. It seems plausible that one's values and one's lived experiences often are connected, or at least not independent of each other. Of course, this is a contingent, imperfect empirical correlation. People with similar lived experiences and social locations may very well have different values. As an example, take two homosexual persons growing up in a heteronormative society. They will likely share experiences that will make them acutely aware of heteronormative behavioural norms, family models, and mechanisms of discrimination. These experiences have the potential to make them question common heteronormative values and to reject any kind of discrimination based on sexual orientation. This need not be so, however; in particular, because people often internalise prejudices against themselves. When it comes to value questions such as whether homosexual couples should be allowed to adopt children, one of our two persons might be in favour – and the other one against, if sharing a common societal belief that children do best with a binary, monogamous, heterosexual couple as parents. It is also perfectly possible that any heterosexual person will adopt egalitarian, anti-discriminatory values. Yet the likelihood of diverging from a heteronormative value consensus will be increased by experiences that don't fit in well with this consensus because of the resulting friction.

If this is so, then a community characterised by diversity in values will, on average, also have a greater degree of situational diversity – and the other way around. In fact, one of the key developments in standpoint theory has been the more careful formulation of the epistemic advantage thesis as requiring a standpoint, which does not come automatically with a social location (see Section 4.1.2). If a standpoint is what provides a potential epistemic advantage, and a standpoint essentially involves critical reflection of one's social experiences and their relation to power structures, it seems likely that those holding a standpoint will also share certain values. For example, developing a critical consciousness for how one's social experience, as well as (scientific) knowledge, is shaped by power structures will likely correlate with a devaluation of oppression and a valuation of equality.

In conclusion, the first disagreement between FST and CCE could be resolved by understanding social diversity both in terms of values and social locations. A fruitful avenue for future research would then lie in the further exploration of this relation between experiences and values, as well as its normative and institutional implications.

From an institutional perspective, a pressing question lies in how to structure and organise scientific research in a way that fosters epistemically beneficial diversity. Fehr (2011) points out that contemporary 'diversity promotes excellence' theories (such as CCE) fail to address certain problems arising from imperfect implementations of situational diversity. For example, she argues that the epistemic benefits of situational diversity can be harnessed by a scientific community that is still perpetuating oppressive power structures. Such 'epistemic free-riding' only requires that there are some researchers from marginalised social locations, even if the respective marginalised groups are still severely underrepresented, and the few to be found in science work in precarious positions (e.g., as poorly paid adjuncts with fixed-term contracts rather than as tenured faculty). To avoid such epistemic free-riding, we need to think about how to implement diversity in science without furthering epistemic and economic exploitation.

6.2 The Role of Values

While there remain problems and open questions regarding the optimal implementation of diversity in science, I have argued above that the disagreement within feminist philosophy of science about whether to understand diversity in terms of values or social locations can be overcome by combining those variables, especially as they will sometimes correlate. However, this solution is affected by the second point of disagreement identified by Intemann (2010): the role assigned to values. In CCE, values play an *instrumental* role. They are ascribed a positive epistemic function insofar as they help to identify implicit value-laden background assumptions.

With regard to this function, all values are, in principle, on a par. Arguably, this leads to an issue regarding the proper limitations of diversity. Hicks (2011) has brought forward the so-called 'Nazi problem' for CCE: if the role of values is only instrumental, and if more value diversity is therefore always epistemically beneficial, this entails that we should not just promote feminist values in science but also contrasting ones. The worry here is that a commitment to CCE entails a commitment also to foster sexist or racist values for the sake of diversity. Critical contextual empiricism therefore might end up advancing anti-feminist and anti-egalitarian aims (see the next section for further discussion).

As pointed out in Section 4, CCE and FST are both social epistemologies. Standpoint theorists extend this shared stance into an explicitly political epistemology. For example, Intemann (2010) argues that values in FST do not only have causal, instrumental roles but can function as *(good) reasons* in themselves. For her, it is not about a diversity of values so much as it is about relying on the *right* values. In CCE, values are problematic insofar as they go unnoticed; in FST, they are problematic in so far as they lack justification.[25]

[25] It is noteworthy that other feminist standpoint theorists propose positions compatible with an instrumental role of values (e.g., Wylie 2003), as an anonymous reviewer has pointed out. On the other side, feminist empiricism can also be combined with the argument that any gap between theory and data should be filled with the right values (e.g., those promoting human flourishing; Kourany 2010).

> For standpoint theorists, the reason that sexist values and androcentrism are bad for science is not because they are values that give rise to partiality. Rather, the problem is that they are unjustified value judgments. The reason that feminist ethical and political commitments do not lead to problematic bias in the same way is because these value judgments are better supported, or warranted. (Intemann 2010, 793)

Holding feminist values to be legitimate in science because they are the right and better-justified values solves the Nazi-problem: We would not need to tolerate, let alone foster, racist or sexist values in science, as these lack justification. However, this solution presumes a very substantial, and probably controversial, premise: namely, that feminist values can be justified as the right ones. It remains unclear how this justification is to proceed. Intemann argues that feminist empiricists have to agree on this premise, since they, after all, are also feminists supporting feminist values (Intemann 2010). In consequence, we might forge a consensus within feminist philosophy of science by understanding the call for diversity as one for more situational diversity and for an increase in feminist values.

The question is how far such a consensus position could carry. First, many people might doubt that just because people agree on something (e.g., feminist values), it is also right or justified. This is important because, second, the relevance of this discussion goes beyond the field of feminist philosophy of science as such, since both CCE and FST aim to provide alternatives to the value-free ideal – for everyone, not just for feminist scholars. Ideally, a feminist consensus position would therefore also serve to provide a general normative alternative to the VFI. The value-free ideal, in turn, serves several functions: enabling the pursuit of truth, providing reliable information for autonomous decision-making, and warranting public trust in science.

It seems likely that those uncommitted to feminist values will dismiss scientific information based on the purportedly right values, and consider it to be biased rather than trustworthy. One might also argue that a democracy actually needs to tolerate a larger degree of dissent regarding value questions. This does not have to mean embracing racist or sexist values; it could amount to more benign differences. For example, not everyone will agree that the promotion of social justice (as opposed to, e.g., economic growth) should be a primary aim of science – or social justice might be understood differently. Others might disagree because they hold non-cognitivist metaethical positions. Relatedly, feminist values can also be interpreted in different ways; just imagine a Marxist ecofeminist versus a liberal feminist. A shared self-identification as feminist leaves room for substantial disagreements.[26] To replace the ideal of

[26] This problem also affects positions that consider feminist values to be neither purely instrumental nor better justified, but foundational commitments that are not in need of justification (cf.

value-freedom with one of committing to the right values, then, requires showing how we can justify what the right feminist values are, and it requires solving the issue of how to ensure autonomy and trustworthiness among those parts of the public that diverge in their value orientations.

The burden of proof here lies on the side of those who want to designate certain feminist values as the right ones. Gesturing at a general agreement about valuing gender equality among feminists will not provide sufficient justification for such a controversial claim. Intemann and de Melo-Martin (2016) inquire further into this issue in the context of biomedical research. They argue that commercial values (i.e., economic profit) are partial in a problematic way, whereas feminist values (i.e., overcoming oppression and power inequalities) are not, for two reasons.

For one, they hold that feminist values are less likely than commercial values to conflict with epistemic aims:

> Attempting to produce research guided by feminist values that simply appears to be effective but (sic) in fact is not would be of little use to feminists. Research that merely appears to be objective is unlikely to combat oppression, help the marginalised or address public health concerns. However, as research funded by tobacco companies, for instance, has shown, that is certainly not the case for research that is influenced by commercial values (Intemann and de Melo-Martin 2016, 83 f).

In this context of medical research and public health, the empirical track-record of research done guided by commercial versus feminist interests certainly speaks in favour of their hypothesis. The question, however, is whether this is due to feminist values being the right ones. For example, the contrast here also hinges on the institutional and political context of research. We might imagine a system in which incentives and disincentives are structured such that only actually safe, effective drugs will generate profits (especially in the mid- and long-term). In line with this, we could also interpret the track-record of feminist vs profit-oriented research to support feminist values in an instrumental sense: that is, as justified insofar as they help promote epistemic aims.

For another, Intemann and de Melo-Martin (2016) hold that feminist but not commercial values are *morally justified* because they do not unfairly prioritise

Hicks 2023; Yap 2016). If feminist values, or the interpretation of these, can conflict, it raises the question of what the right basic commitments are. For instance, Hicks (2011) uses the example of the Islamic feminist al-Faruqi, who interprets feminism in a way compatible with traditional gender roles (providing care work versus economic stability) and differentiates between equality and identity (of rights and responsibilities). Al-Faruqi rightly notes how her account is at odds with the individualism inherent in much of Western feminism (al-Faruqi 1983). It also is fundamentally at odds with premises shared widely in feminist philosophy of science, such as considering equal gender representation in academic careers a desirable goal.

the interests of some stakeholders over others. Assuming that scientific knowledge should contribute to the common good, they argue that the interests of marginalised groups should be prioritised because they suffer disproportionately from avoidable health issues. While as a feminist, I strongly agree with this, it does *presume* rather than justify egalitarian, feminist values, and thus might not convince the neoliberal capitalist who values profit over equality.

Another route to the justification of values has been proposed by *radical feminist empiricists*, who argue that value judgements can be supported by empirical evidence. Starting from the problem of underdetermination and Quinean holism, it is proposed that the holistic net contains not only empirical but also conventional and evaluative judgements – including non-epistemic value judgements. Solomon (2012) has coined the term 'web of valief' for this inclusive version of holism, in which values would be subject to empirical testing in conjunction with the theoretical hypothesis for which they serve as background assumptions (e.g., Anderson 2004; Clough 2004; Nelson 1990).

As Anderson (2004) argues, value judgements cannot be logically entailed by empirical judgements – but neither can any theoretical statements that go beyond empirical observations. Yet it is often concluded from this that value-judgements (unlike theoretical statements) cannot stand in other evidentiary relations to empirical evidence either. Anderson points out that value-judgements often have, or relate to, descriptive content. For example, a value-judgement that women should not work in STEM fields might be supported by a belief that women are bad at maths; such a value-judgement can therefore conflict with empirical evidence showing this belief to be mistaken. This implies that one sometimes has to update one's value-judgements in response to scientific empirical research if one wants those value-judgements to be rational. The problem regarding rationality, Anderson argues, stems not from the evaluative content of value-judgements, but from holding them dogmatically. According to this, not only is science not value-free – but values, too, are not, or should not, be science-free.

Anderson is right to point out the mutual entanglement of facts and values and accompanies her argument by a case study on interactions between empirical results and normative beliefs in research on divorce. However, feminist radical empiricism is not yet fully developed. For instance, it remains unclear whether we can generalise from the divorce case study (Solomon 2012), and how we can specify the evidentiary relations between values and facts (Yap 2016). This presents an interesting starting point for further research in light of the under-theorisation of value judgements in the current debate. For now, though, it seems a stretch to argue that (a) values can be justified empirically, and (b) (specific) feminist values have been justified in that manner across all specific contexts.

Considering this, feminist standpoint empiricism might be argued to be better served by agreeing (for now) on an instrumental role for feminist values, rather than relying on their better justification. As noted above, this is at least compatible with some accounts in FST. Intemann refuses this, however, because it runs into the Nazi-problem, as sketched above. In the next section, I will discuss whether the Nazi-problem can be solved holding on to a solely instrumental role for values.

6.3 The Nazi-Problem in Feminist Empiricism

As mentioned above, Dan Hicks has introduced the Nazi-problem as a consequence of CCE calling for an overall diversity of values, rather than preferencing feminist values. Rolin (2017) argues that CCE has the resources to get rid of the Nazi, as sexist or racist ideologies conflict with central tenets of social objectivity. In particular, this holds for the requirement of equality of intellectual authority (tempered only in relation to expertise, not sociodemographic variables) and the requirement of uptake of criticism (from a variety of perspectives). Nazi scientists are considered unlikely to comply with these requirements, as they go against Nazi ideology. However, Hicks (2011) worries that, as long as the Nazi scientists are in the minority, they might pay lip service to the norms of social objectivity, while in fact undermining the transformative critical process – never violating but constantly challenging the rules.

This leads Hicks to discuss a potential solution, namely that a Nazi could not participate in a respective critical process and diverse community *in good faith*. Hicks identifies two central values underlying Longino's social value management ideal: (a) *formal egalitarianism*, implying that all members of the community should have the same formal standing, regardless of race, gender, class, and so on, and (b) *liberal pluralism*, which allows for reasonable disagreement; that is, a conflict of positions does not imply that one of the positions is irrational. These conflict with fundamental ethico-political values of Nazi ideology (Hicks 2011, 340ff.).

Adding a good faith requirement to CCE can solve the Nazi-problem: we do not need to foster perspectives that stand in tension with its underlying values. According to Hicks (2011) and Intemann (2010, 2017) it does not, however, solve the problem of CCE failing to be feminist. This is so because, arguably, the good faith requirement cannot be met by certain feminist perspectives either. We face a dilemma here: CCE without the good faith requirement is overly inclusive; CCE with the good faith requirement is overly exclusive.

Hicks discusses two families of feminism, which they argue stand in conflict with liberal pluralism and formal egalitarianism. First, this concerns various kinds of communitarian feminist positions (e.g., Islamic or Christian feminisms). Some of these deny formal egalitarianism, trying to reconcile feminism with differential rights and responsibilities for people of different sex/gender (cf. footnote 26). Others are incompatible with liberal pluralism insofar as they commit to the ideal of a community sharing a 'comprehensive conception of the good human life'. Rejections of fundamental elements of this conception of the good will, by the standards of the respective community, appear unreasonable (Hicks 2011, 346ff.).

In response, I think CCE could bite the bullet here. For one, denying equal formal standing to people of different genders (just think about what this would sound like expanded to different sexual orientations, races, or levels of ability) might rightly serve to disqualify a perspective from the transformative critical process, even if it calls itself feminist. For another, one could argue that casting perspectives that diverge from one's own as unreasonable or irrational embodies a dogmatism that is, in fact, problematic in scientific discourse. This is not just about believing one's own values to be right, but about denying that conflicts on this level allow for rational discussion.

The second strand of feminism that Hicks argues is excluded by the good faith criterion is feminist standpoint theory. The problem here arises because of the epistemic advantage thesis in FST, which implies that those with feminist standpoints (politically reflected social locations) should be given greater epistemic privileges than other social groups. Hicks holds this to conflict with formal egalitarianism (Hicks 2011, 345–46).

As FST has been one of the most influential approaches within feminist philosophy of science, feminist empiricists will want to dodge this bullet. A possible response could be to question whether this conflict needs to arise on all interpretations of FST. For one, many standpoint theorists understand the connection between social location and standpoint as contingent, contextual, and non-automatic (see chapter 4). That is, FST does not necessarily entail the claim that researchers with certain socio-demographic characteristics should be given a higher privilege or standing in the scientific community than others solely by virtue of their socio-demographics. For another, the value of formal egalitarianism underlies the requirement for tempered equality of intellectual authority in CCE. This holds that authority should not be distributed differentially according to any other factors than intellectual merit. It thus cannot give someone more authority simply based on their gender, for example. Yet, if we understand a standpoint as politically reflected lived experience, it could arguably fulfil the criteria for rightfully tempering authority: Certain marginalised standpoints can, in particular contexts, be expected to be particularly epistemically fruitful and *therefore* deserving of higher than usual intellectual authority.

To sum this up, I think one can advocate a *feminist standpoint empiricism* that, in contrast to Intemann's proposal, leans more heavily onto the empiricist side. It could combine a diversity of social locations and values, while ascribing an instrumental role to the latter. The ensuing Nazi-problem could be solved by adding the good faith requirement, which, in turn, would exclude communitarian feminists but needs not be interpreted in a way that excludes feminist standpoint theorists.

Calling for a diversity of values, however, does not mean that we should treat all these different value-laden perspectives equally, either. Instead, it seems important to integrate some insights of standpoint theory: values and voices of non-dominant groups deserve preferential treatment. One can justify this on instrumental grounds by arguing that preferential treatment would increase epistemic injustice in scientific communities – which, in turn, is needed to reap the benefits of diversity.

6.4 Epistemic Injustice in Science

Epistemic injustice is an ethical-epistemological concept designating the wrong of undermining someone in their capacity as a knower. In her seminal book, Fricker (2007) distinguishes two main kinds of epistemic injustice: testimonial and hermeneutical injustice. *Testimonial injustice* pertains to cases in which a speaker is assigned an inadequately low degree of credibility by a hearer owing to negative identity prejudices related to either competence or sincerity. For instance, someone might disbelieve a female speaker's explanation of how quantum entanglement works due to social prejudices (e.g., the belief that women aren't good at matters of science and technology). Relevant prejudices can be localised or systematic and persistent, in which case they track a speaker throughout all their social life.

Such testimonial injustices are harmful in a variety of ways. First, they are ethically unfair and degrade someone in their capacity as a knower, a capacity that we hold to be essential to our very humanity. Systematic and persistent testimonial injustice also brings with it a lot of potential follow-on harms, such as career disadvantages, poverty, incarceration, and a lack of epistemic self-trust (Fricker 2007, chapter 2). Second, testimonial injustices are epistemically harmful in that they constitute dysfunctions in testimonial exchange, thus hindering the successful transmission of knowledge. As testimonial injustice affects certain social groups disproportionately, this will foster ignorance with regard to certain kinds and areas of (scientific) knowledge. If we take seriously the ideas that knowledge is socially situated, and marginalised social locations are often blocked from successful testimonial exchange, the shared knowledge pool ends up severely and systematically diminished.

Matters get even worse if we think of testimonial injustice in broader terms of *participatory injustice.* Hookway (2010) distinguishes an *informational* from a *participant perspective* on epistemic injustice. The former captures the traditional sense of testimony as communicating knowledge in the form of assertions, which can fail due to prejudice. Yet our epistemic practices range far beyond the transmission of information, particularly when it comes to the generation of new (scientific) knowledge. This involves asking critical questions, coming up with new ideas to try out, thinking of examples and counterexamples, playing devil's advocate for a contested position, or having any kind of discussion. With regard to these practices, participatory epistemic injustice can occur when I do not consider somebody a fully capable epistemic contributor (even when I trust them to correctly report about existing knowledge). One of Hookway's examples is the misunderstanding of a critical question as one of clarification: that is, as located on the informational versus participatory plane.

Hermeneutical injustice is defined by Fricker as 'the injustice of having some significant area of one's social experience obscured from collective understanding owing to hermeneutical marginalization' (Fricker 2007, 158). Hermeneutic marginalisation refers to the exclusion from (or underrepresentation in) meaning-making practices, such as law, art, science, philosophy, or journalism. These hermeneutical practices shape our constantly evolving pool of shared conceptual resources, which allows us to make sense of our experiences and to communicate them to other people. If people belonging to certain social groups are underrepresented here, the resulting pool of conceptual resources can lack the means for interpreting their experiences. One of Fricker's examples is the experience of sexual harassment *before* the concept existed, which made it difficult for the harassed person to understand their own experience and to talk about it in ways that get the point across. Fricker conceives of hermeneutical injustice in terms of general conceptual lacunae that affect both the harassed and the harasser. However, the disadvantage resulting from this shared lack is asymmetrical, in that its consequences are worse for the person being harassed while being unable to articulate this experience effectively (Fricker 2007, chapter 7).

Other authors have developed the concept of epistemic injustice further and have identified additional forms of it (cf. also Fricker 2017 for a response). The concept of hermeneutical injustice has been expanded to include cases in which a marginalised community successfully creates resources to make sense of their social experiences, yet there is no uptake by the speakers belonging to dominant groups (what Dotson 2011 calls *contributory injustice*), or these resources are wilfully ignored (Pohlhaus jr. 2012). An example relevant to science might be the neurodiversity paradigm developed by people living with certain atypical traits traditionally conceptualised as mental disorders (such as autism or schizophrenia). Neurodiversity is opposed to the pathology paradigm central to much of psychiatric research and clinical practice (see also Knox 2022). Other cases would include the initial dismissal of feminist research in a certain field (e.g., primatology, medicine, or evolutionary biology) as irrelevant.

Epistemic injustice in all its different forms plays a very problematic role in science (cf. also Grasswick 2017). For example, scientists belonging to marginalised groups can repeatedly experience not being believed when making assertions about their area of competence, or of being treated differentially when it comes to participation in the many epistemic activities involved in scientific research. It also has clear, quantifiable effects that can shape (or even end) scientific careers. Basically, science is a prestige economy centring on epistemic credibility. If one's credibility is systematically and persistently underrated, the effects are far-reaching, as it influences important variables such as citation counts, opportunities for collaboration and co-authorships, and research grants.

Beyond the harms that systematic unfair credibility deflation can have for individuals belonging to socially marginalised groups, it should be clear from the arguments and examples presented in this Element that the consequences in terms of the knowledge produced (or not produced) are also extensive. From the perspective of standpoint epistemology, discrediting or ignoring input from members of marginalised groups deprives scientific communities of precisely those perspectives that are connected to social locations with potential epistemic advantages. As Grasswick (2017) points out, the various forms of epistemic injustice also present obstacles to realising the conditions for social objectivity in Longino's account. As mentioned above, these conditions include the uptake of criticism and tempered equality of intellectual authority. If there is a systematic bias in whose criticism is considered relevant and is responded to, and who is taken to be capable of criticising background assumptions, objectivity is diminished.

The concept of epistemic injustice can therefore help to further spell out the idea of epistemically beneficial diversity. It will not be enough to have diversity in the sense of fair representation of different social groups of science. If representation continues to be accompanied by credibility deflations, the social objectivity of a community's research process will still suffer. For instance, if the percentage of female primatologists rises from 5 to 50 per cent but a critical feminist perspective on mainstream primatological research is still belittled, work by female researchers is still under-cited, and female researchers still have to fight harder for recognition than their male counterparts, even equal representation is unlikely to yield the epistemic benefits of diversity.

The envisaged epistemic benefits of diversity require people to point out blind spots and implicit value-laden background assumptions. Much of the needed epistemic diversity work thus consists in critiquing existing research and questioning consensus positions, as well as providing alternative approaches for further development. To successfully do this, researchers need not only strong skills of critical reflection and innovative ideas; they also need to be heard and taken seriously, to be supported in the early phases of a research project, and so on. Thus, for diversity to be epistemically beneficial, credibility assessments need to be fair and adequate, as the criterion of equality in intellectual authority points out.

What is problematic about this is how scientific credibility assessments are motivated or justified. Often, such assessments are based on people's position in the academic hierarchy and their given standing in a certain field (involving indicators such as publications, citations, grants, or keynotes). Even if such assessments and indicators were free from social biases, credibility was still tied to prior success. Such success will, to a greater or lesser extent, require one to fit in with certain research traditions, practices, and standards relevant to a specification of what 'good' science is. For example, this includes adherence to certain epistemic values, but it is much broader in that it also connects to norms of behaviour and habitus that are often gender-, race-, or class-related.[27]

Repeated experiences of epistemic injustices can lead to forms of (self-) silencing: either people might refrain from expressing criticism and new ideas altogether, or they might shape their utterances in a way that ensures uptake by catering to listener's expectations (Dotson 2011). A researcher from a marginalised group is put in a very difficult position here, both from a psychological and a prudential perspective. To build an academic career, they need to demonstrate excellence that is recognised as such, and will therefore often need to be congruent with current ideas of what good science looks like. To further the epistemic benefits of diversity, however, they would need to be explicitly critical about assumptions, practices, and standards informing the idea of good science and the state of the art in their field. These are conflicting demands. Epistemic diversity work, then, might lead to one's

[27] Rolin (2002) makes a strong case for the gendered nature of credibility assessments. She distinguishes between a scientist's actual and perceived trustworthiness (i.e., credibility). The latter is affected by behavioural norms in the sense that stereotypical masculine behaviours and traits, such as assertiveness and competitiveness, increase credibility. Men, therefore, can do science while doing gender, whereas women cannot, and face conflicting behavioural norms.

marginalisation within one's research field. This way, it will be difficult to gain the credibility (as based on indicators of academic success) needed to get sufficient uptake for one's critique in the first place.

In consequence, diversity of social location and/or values as such is insufficient for epistemically beneficial diversity. Such diversity needs to be embedded in an epistemically just credibility economy in the respective scientific communities. It also requires more than equal distribution of credibility among researchers. The credibility economy needs to be such that epistemic diversity work can correct for former injustices and resulting biases. As mentioned above, this also needs to be organised in a way that does not lead to the epistemic exploitation of already socially marginalised researchers. Critical and divergent work from someone in a marginalised social position would need to *enhance*, rather than diminish, their credibility.

Fricker (2007) embeds her account of epistemic injustice in a framework of virtue ethics. The idea is that epistemic injustices should be countered by the exercise of the virtues of testimonial and hermeneutical justice. These virtues require hearers to correct for inadequate, prejudiced credibility assessments. To do so, hearers need to be aware of their own and the other persons' social location, as well as how these social locations relate to prejudices, and of their potential dysfunctional results in terms of knowledge exchange and production. Given such awareness, one can correct for prejudice by assigning someone higher credibility than one otherwise would. Talking to, or inquiring with, people who belong to marginalised groups, one should, so to say, 'round up' their credibility level and give them the benefit of the doubt when what they say seems implausible or hard to understand.

To counter epistemic injustice in science, and thereby enable fruitful and worthwhile epistemic diversity work, scientists could thus train themselves in these virtues of epistemic justice. Beyond that, it is important also to think about the issue of a credibility economy on a structural and institutional level. For example, this applies to the usage of indicators of scientific excellence (e.g., one's h-index or the impact factor of the journals one has published in), now frequently used in employment and grant decisions, or the preference for original over critical work. How exactly to implement and measure an epistemically just and beneficial credibility economy on an institutional level is an important question for further research.

In general, integrating a focus on epistemic justice could help with achieving some kind of feminist empiricist standpoint consensus position in the field. As has been argued above, such a position should aim for a diversity of social location and values due to their instrumentally beneficial role, rather than for shared commitment to the right feminist values. Yet this does not mean that all values are created equal: values that conflict with participating in an open critical discourse in good faith can be excluded. If, as I have argued, epistemically beneficial diversity requires epistemic justice, then value orientations that conflict with the virtues of testimonial and hermeneutic justice could also be ruled out: the sexist scientist is very unlikely to assign more, rather than less, credibility to researchers who are gendered female. The dreaded Nazi scientist likewise will struggle with giving more credibility to researchers of colour owing to these researchers' first-hand experiences of racism.

Here again, the need for a diversity in both values and social locations arises. Epistemic injustices work by way of prejudices against certain social groups, and the assumption is that our hermeneutical resources are shaped by our lived experiences and our attempts to make sense of these. To involve people with different lived experiences is thus crucial; a point made most strongly by standpoint theorists. Last but not least, standpoint theory can add a highly needed emphasis on issues of agenda-setting to the feminist empiricist position. Whereas CCE focuses mostly on the vetting of theories in a critical process, FST widens the focus to what questions are actually asked and whose interests have so far been ignored, thereby increasing the chance of overcoming former blind spots. This also puts a spotlight on the study of scientific ignorance as a highly relevant line of research to further explore.

7 Conclusion and Future Directions

Science is not just politics by other means – but its organisation, as well as the resulting knowledge, are deeply political nevertheless. Feminist philosophy of science has done much to show these interconnections and has generated a rich pool of ideas about how to make science better, both epistemically and politically. Science was and is negatively affected by social oppression, which diminishes epistemically beneficial diversity. Feminist philosophers of science have successfully moved these pressing issues about the relations between science and society from the philosophical margins to the centre stage.

In this Element, I have mainly focused on relations between science and gender. However, since feminist philosophy of science is concerned with the epistemic effects of oppression and the effects of scientific knowledge-production on social injustices, it has been developing into a more broadly conceived critical and liberatory epistemology. This follows from its emphasis on the role of lived experience – situated knowledge is differential knowledge. Gender is, of course, not the only variable relevant to social power hierarchies and marginalisation; and the power matrix is further differentiated by intersectionality. Experiences of discrimination and social disadvantages – and thereby any potential epistemic advantages – will vary substantially based on other factors such as race or class and their interconnections. Feminist philosophy of science increasingly incorporates this insight, and exciting new work harnesses the connections to related research areas such as the philosophy of disability, decolonialism, critical race theory, or queer studies (cf., e.g., Gupta & Rubin 2020; Harding & Mendoza 2020; Tremain 2020).

There remains work to be done, of course; for example, with regards to the right conceptualization of diversity in science, as well as how best to institutionalise it. In the previous chapter, I pointed to how the burgeoning literature on epistemic injustice can provide new ways of thinking about scientific communities, and about what the interactive process of creating scientific knowledge should optimally look like. Additional issues arise from the fact that social marginalisation often involves a lack of access to education, especially the kind of resource-intensive education required for an academic career. Arguably, diversity in scientific communities could thus always turn out to be too limited, in the sense of excluding exactly those perspectives or standpoints that are the most marginalised. One potential way to counter this problem is to increase the amount of community-based and participatory research that actively integrates lay perspectives (e.g., Grasswick 2010, 2017; Whyte and Crease 2010).

As science develops in an ever-changing world, new challenges will also continue to arise with regards to scientific and technological innovations that need to be reflected from a feminist perspective (such as, e.g., the increased use of algorithmic decision-making in various social contexts). Moreover, I have argued that feminist critics of the value-free ideal have successfully discredited it and have provided alternatives that do not threaten the epistemic integrity of science. This has also pointed to questions that could be fruitful for further research, for example, about the relation between values and social situatedness, whether there can or even should be a feminist consensus position, or how to conceptualise the relation between value judgements and empirical judgements.

Other crucial questions for future research lie in exploring how feminist alternatives to the value-free ideal can also ensure, or even enhance, autonomous decision-making and public trust in science-based policy. Feminist philosophers have argued that the ultimate goal for objectivity is not so much correct representation as, rather, enabling public trust – and they argue that this, too, is an issue affected by social situatedness (Grasswick 2010; Scheeman 2001). It depends very much on who the knowledge is produced by and for, and how it is communicated. Philosophical work on trust in science has been flourishing recently, and will likely continue to do so.

The last years have made it brutally clear what huge challenges humanity is facing, ranging from climate change, to pandemics, to wars, to concerted political efforts to destabilise democracies. The need for scientific knowledge and its wise and fair usage seems greater than ever, at a time when political debate is increasingly polarised and misinformation abundant. Understanding and shaping the relations between science and society is one of the most pressing issues in the twenty-first century. It is one about which feminist philosophers of science have much to say.

References

al-Faruqi, L. L. (1983). Islamic traditions and the feminist movement: Confrontation and cooperation. *The Islamic Quarterly*, *27*(3), 135–39.

Altmann, J. (1974). Observational study of behavior: Sampling methods. *Behaviour*, *49*(3–4), 227–66.

Anderson, E. (1995). Feminist epistemology: An interpretation and a defense. *Hypatia*, *10*(3), 50–84.

Anderson, E. (2004). Uses of value judgments in science: A general argument, with lessons from a case study of feminist research on divorce. *Hypatia*, *19*(1), 1–24.

Anderson, E. (2006). How not to criticise feminist epistemology: A review of *Scrutinising Feminist Epistemology*. www-personal.umich.edu/~eandersn/hownotreview.html.

Anderson, E. (2020). Feminist Epistemology and Philosophy of Science. In E. N. Zalta, ed., *The Stanford Encyclopedia of Philosophy* (Spring 2020 Edition), https://plato.stanford.edu/archives/spr2020/entries/feminism-epistemology/.

Antony, L. (1993). Quine as feminist: The radical import of naturalised epistemology. In L. Antony and C. Witt, eds., *A Mind of One's Own: Feminist Essays on Reason and Objectivity*. Boulder: Westview, pp.110–53. (2nd edition, 2018).

Antony, L. (1995). Sisters, please, I'd rather do it myself: A defense of individualism in feminist epistemology. *Philosophical Topics*, *23*(2), 59–94.

Banco, D., Chang, J., Talmor, N., et al. (2022). Sex and race differences in the evaluation and treatment of young adults presenting to the emergency department with chest pain. *Journal of the American Heart Association*, *11*(10), e024199.

Bateman, A. (1948). Intra-sexual selection in Drosophila. *Heredity*, *2*(3), 349–68.

Belenky, M. F., Clinchy, B. M., Goldberger, N. R., & Tarule, J. M. (1986). *Women's Ways of Knowing: The Development of Self, Voice and Mind*. New York: Basic Books.

Betz, G. (2013). In defence of the value free ideal. *European Journal for Philosophy of Science*, *3*, 207–20.

Biddle, J. (2013). State of the field: Transient underdetermination and values in science. *Studies in History and Philosophy of Science Part A*, *44*(1), 124–33.

Biddle, J., & Leuschner, A. (2015). Climate skepticism and the manufacture of doubt: Can dissent in science be epistemically detrimental? *European Journal for Philosophy of Science*, *5*, 261–78.

Bivins, R. (2000). Sex cells: Gender and the language of bacterial genetics. *Journal of the History of Biology*, *33*, 113–39.

Bleier, R. (1984). *Science and Gender: A Critique of Biology and Its Theories on Women*. Elmsford: Pergamon Press.

Bluhm, R. (2020). Neurosexim and our understanding of sex differences in the brain. In S. Crasnow and K. Intemann, eds., *The Routledge Handbook of Feminist Philosophy of Science*. New York: Routledge, pp. 316–27.

Boston Women's Health Book Collective (1973/1976). *Our Bodies, Ourselves: A Book by and for Women* (2nd rev. ed.). New York: Simon and Schuster.

Botero, M. (2020). Observing primates: Gender, power, and knowledge in primatology. In S. Crasnow and K. Intemann, eds., *The Routledge Handbook of Feminist Philosophy of Science*. New York: Routledge, pp. 275–88.

Bright, L. K. (2018). Du Bois' democratic defence of the value free ideal. *Synthese*, *195*(5), 2227–45.

Bueter, A. (2010). Social objectivity and the problem of local epistemologies. *Analyse & Kritik*, *32*(2), 213–30.

Bueter, A. (2012). *Das Wertfreiheitsideal in der sozialen Erkenntnistheorie*. Frankfurt: Ontos.

Bueter, A. (2015). The irreducibility of value-freedom to theory assessment. *Studies in History and Philosophy of Science Part A*, *49*, 18–26.

Bueter, A. (2017). Androcentrism, feminism, and pluralism in medicine. *Topoi*, *36*(3), 521–30.

Bueter, A. (2022). Bias as an epistemic notion. *Studies in History and Philosophy of Science*, *91*, 307–15.

Bueter, A., & Jukola, S. (2020). Sex, drugs, and how to deal with criticism: The case of Flibanserin. In A. LaCaze and B. Osimani, eds., *Uncertainty in Pharmacology*. Cham: Springer, pp. 451–70.

Burke, M. A., & Eichler, M. (2006). The BIAS FREE framework: A practical tool for identifying and eliminating social biases in health research. *Global Forum for Health Research*.

Campbell, R. (1998). *Illusions of Paradox: A Feminist Epistemology Naturalised*. Lanham, MD: Rowman & Littlefield.

Carrier, M. (2013). Values and objectivity in science: Value-ladenness, pluralism and the epistemic attitude. *Science & Education*, *22*, 2547–68.

Ceci, S. J., & Williams, W. M. (2011). Understanding current causes of women's underrepresentation in science. *Proceedings of the National Academy of Sciences*, *108*(8), 3157–62.

Chatterjee, P., & Werner, R. M. (2021). Gender disparity in citations in high-impact journal articles. *JAMA Network Open*, *4*(7), e2114509–e2114509.

ChoGlueck, C. (2022). Still no pill for men? Double standards & demarcating values in biomedical research. *Studies in History and Philosophy of Science Part A*, *91*, 66–76.

Clarke, E. (1873/2007). *Sex in Education, or: A Fair Chance for Girls*. Rockville: Wildside, .

Clough, S. (2003). *Beyond Epistemology: A Pragmatist Approach to Feminist Science Studies*. Lanham: Rowman and Littlefield.

Clough, S. (2004). Having it all: Naturalised normativity in feminist science studies. *Hypatia*, *19*(1), 102–18.

Clune-Taylor, C. (2020). Is sex socially constructed? In S. Crasnow and K. Intemann, eds., *The Routledge Handbook of Feminist Philosophy of Science*. New York: Routledge, pp. 187–200.

Code, L. (1991). *What Can She Know? Feminist Theory and the Construction of Knowledge*. Ithaca: Cornell University Press.

Cole, J. R. (1987). *Fair Science: Women in the Scientific Community*. Repr. New York: Columbia University Press.

Crasnow, S. (2020). Feminist Perspectives on Science, In E. N. Zalta, ed., *The Stanford Encyclopedia of Philosophy* (Winter 2020 Edition), https://plato.stanford.edu/archives/win2020/entries/feminist-science/.

De Melo-Martin, I. (2020). The gendered nature of reprogenetic technologies. In S. Crasnow and K. Intemann, eds., *The Routledge Handbook of Feminist Philosophy of Science*. New York: Routledge, pp. 289–99.

Dotson, K. (2011). Tracking epistemic violence, tracking practices of silencing. *Hypatia*, *26*(2), 236–57.

Douglas, H. (2000). Inductive risk and values in science. *Philosophy of Science*, *67*(4), 559–79.

Douglas, H. (2004). The irreducible complexity of objectivity. *Synthese*, *138*, 453–73.

Douglas, H. (2009). *Science, Policy, and the Value-Free Ideal*. Pittsburgh: University of Pittsburgh Press.

Dreifus, C., ed. (1978). *Seising Our Bodies: The Politics of Women's Health*. New York: Vintage Books.

Duhem, P. M. (1906/1954). *The Aim and Structure of Physical Theory*. Princeton: Princeton University Press.

Dusek, V. (2006). *Philosophy of Technology: An Introduction*. Oxford: Blackwell.

Dworkin, J. D., Linn, K. A., Teich, E. G., et al. (2020). The extent and drivers of gender imbalance in neuroscience reference lists. *Nature Neuroscience*, *23*(8), 918–26.

Ehrenreich, B., & English, D. (1978). *For Her Own Good: 150 Years of the Experts' Advice to Women*. New York: Anchor Books.

Eichler, M. (1988). *Non-sexist Research Methods: A Practical Guide*. New York: Routledge.

Elliott, K. C. (2022). *Values in Science*. Cambridge: Cambridge University Press.

Elliott, K. C., & McKaughan, D. J. (2009). How values in scientific discovery and pursuit alter theory appraisal. *Philosophy of Science*, *76*(5), 598–611.

Epstein, S. (2007). *Inclusion: The Politics of Difference in Medical Research*. Chicago: University of Chicago Press.

European Commission, Directorate-General for Research and Innovation (2021). *She figures 2021 – Gender in research and innovation – Statistics and indicators*, Publications Office, https://data.europa.eu/doi/10.2777/06090.

Fausto-Sterling, A. (1985). *Myths of Gender: Biological Theories about Women and Men*. New York: Basic Books.

Fedigan, L. M. (1992). *Primate Paradigms: Sex Roles and Social Bonds*. Chicago: University of Chicago Press.

Fedigan, L. M. (1997). Is primatology a feminist science?. In L. D. Hager, ed., *Women in Human Evolution*. London: Routledge, pp. 55–74.

Fehr, C. (2011). What is in it for me? The benefits of diversity in scientific communities. In H. E. Grasswick, ed., *Feminist Epistemology and Philosophy of Science: Power in Knowledge*. Dordrecht: Springer, pp. 133–55.

Fehr, C. (2018). Feminist philosophy of biology. In E. N. Zalta, ed., *The Stanford Encyclopedia of Philosophy* (Fall 2018 Edition), https://plato.stanford.edu/archives/fall2018/entries/feminist-philosophy-biology/.

Fernández Pinto, M. (2020). Ignorance, science, and feminism. In S. Crasnow and K. Intemann, eds., *The Routledge Handbook of Feminist Philosophy of Science*. New York: Routledge, pp. 225–35.

Fischer-Homberger, E. (1979). *Krankheit Frau und andere Arbeiten zur Medisingeschichte der Frau*. Bern: Hans Huber.

Fox Keller, E. (1983). *A Feeling for the Organism: The Life and Work of Barbara McClintock*. San Francisco: W.H. Freeman.

Fricker, M. (2007). *Epistemic Injustice: Power and the Ethics of Knowing*. Oxford: Oxford University Press.

Fricker, M. (2017). Evolving concepts of epistemic injustice. In I. J. Kidd, J. Medina, and G. Pohlhaus Jr., eds., *The Routledge Handbook of Epistemic Injustice*. London: Routledge, pp. 53–60.

Gagné-Julien, A. (2021). Wrongful medicalization and epistemic injustice in psychiatry: The case of premenstrual dysphoric disorder. *European Journal of Analytic Philosophy*, *17*(2), 4–36.

Gilligan, C. (1982). *In a Different Voice: Psychological Theory and Women's Development*. Cambridge, MA: Harvard University Press.

Godfrey-Smith, P. (2009). *Theory and Reality: An Introduction to the Philosophy of Science*. Chicago: University of Chicago Press.

Grasswick, H. E. (2004). Individuals-in-communities: The search for a feminist model of epistemic subjects. *Hypatia*, *19*(3), 85–120.

Grasswick, H. E. (2010). Scientific and lay communities: Earning epistemic trust through knowledge sharing. *Synthese*, *177*, 387–409.

Grasswick, H. (2017). Epistemic injustice in science. In I. J. Kidd, J. Medina, and G. Pohlhaus Jr., eds., *The Routledge Handbook of Epistemic Injustice*. London: Routledge, pp. 313–23.

Grasswick, H. (2018). Feminist social epistemology. In E. N. Zalta, ed., *The Stanford Encyclopedia of Philosophy* (Fall 2018 Edition), https://plato.stanford.edu/archives/fall2018/entries/feminist-social-epistemology/.

Gupta, K., & Rubin, D. A. (2020). Queer Science Studies/Queer Science. In S. Crasnow and K. Intemann, eds., *The Routledge Handbook of Feminist Philosophy of Science*. New York: Routledge, pp. 131–43.

Haack, S. (1993). Epistemological reflections of an old feminist. *Reason Papers*, *18*, 31–43.

Haack, S. (1996). Science as social? – Yes and no. In L. H. Nelson and J. Nelson, eds., *Feminism, Science, and the Philosophy of Science*. Dordrecht: Kluwer Academic, pp. 79–93.

Haas, R., Watson, J., Buonasera, T., et al. (2020). Female hunters of the early Americas. *Science Advances*, *6*(45), eabd0310.

Hall, R. M., & Sandler, B. R. (1982). *The Classroom Climate: A Chilly One for Women?* Washington, DC: Association of American Colleges.

Hall, R. M., & Sandler, B. R. (1984). *Out of the Classroom: A Chilly Campus Climate for Women?* Washington, DC: Association of American Colleges.

Handelsman, J., Cantor, N., Carnes, M., et al. (2005). More women in science. *Science*, *309*(5738), 1190–91.

Haraway, D. J. (1988). Situated knowledges: The science question in feminism and the privilege of partial perspective. *Feminist Studies*, *14*(3), 575–99.

Haraway, D. J. (1989). *Primate Visions: Gender, Race, and Nature in the World of Modern Science*. New York: Routledge.

Harding, S. (1986). *The Science Question in Feminism*. Ithaca: Cornell University Press.

Harding, S. (1991). *Whose Science? Whose Knowledge? Thinking from Women's Lives*. Ithaca: Cornell University Press.

Harding, S. (1992a). After the neutrality ideal: Science, politics, and 'strong objectivity'. *Social Research*, *59*(3), 567–87.

Harding, S. (1992b). Rethinking standpoint epistemology: What is 'strong objectivity'? *The Centennial Review*, *36*(3), 437–70.

Harding, S. (2015). *Objectivity and Diversity: Another Logic of Scientific Research*. Chicago: University of Chicago Press.

Harding, S., & Mendoza, B. (2020). Latin America decolonial feminist philosophy of knowledge production. In S. Crasnow and K. Intemann, eds., *The Routledge Handbook of Feminist Philosophy of Science*. New York: Routledge, pp. 104–16.

Hicks, D. (2011). Is Longino's conception of objectivity feminist? *Hypatia*, *26*(2), 333–51.

Hicks, D. (2023). Revisiting the Nazi Problem. Blog post, https://dhicks.github.io/posts/2023-06-07-nazi-problem.html.

Hill Collins, P. (1990). *Black Feminist Thought: Knowledge, Consciousness, and the Politics of Empowerment*. New York: Routlege.

Hoffmann, D. E., & Tarzian, A. J. (2001). The girl who cried pain: A bias against women in the treatment of pain. *Journal of Law, Medicine & Ethics*, *29*(1), 13–27.

Holman, B., & Wilholt, T. (2022). The new demarcation problem. *Studies in History and Philosophy of Science*, *91*, 211–20.

Hookway, C. (2010). Some varieties of epistemic injustice: Reflections on Fricker. *Episteme*, *7*(2), 151–63.

Hrdy, S. B. (1986). Empathy, polyandry, and the myth of the coy female. In R. Bleier, ed., *Feminist Approaches to Science*. New York: Pergamon Press, pp. 119–46.

Hrdy, S. B. (1981/2009). *The Woman that Never Evolved*. Cambridge, MA: Harvard University Press.

Hubbard, R. (1983). Have only men evolved? In S. Harding and M. Hintikka, eds., *Discovering Reality: Feminist Perspectives on Epistemology, Metaphysics, Methodology, and Philosophy of Science*. Dordrecht: D. Reidel, 45–70.

Hudson, N. (2022). The missed disease? Endometriosis as an example of 'undone science'. *Reproductive Biomedicine & Society Online*, *14*, 20–27.

Hundleby, C. (2020). Thinking outside-in. In S. Crasnow and K. Intemann, eds., *The Routledge Handbook of Feminist Philosophy of Science*. New York: Routledge, pp. 89–103.

Intemann, K. (2005). Feminism, underdetermination, and values in science. *Philosophy of Science*, *72*(5), 1001–12.

Intemann, K. (2010). 25 years of feminist empiricism and standpoint theory: Where are we now? *Hypatia*, *25*(4), 778–96.

Intemann, K. (2017). Feminism, values, and the bias paradox: Why value management is not sufficient. In K. C. Elliott and D. Steel, eds., *Current Controversies in Values and Science*. New York: Routledge, pp. 130–44.

Intemann, K., & de Melo-Martin, I. (2016). Feminist values, commercial values, and the bias paradox in biomedical research. In M. C. Amoretti and N. Vassallo, eds., *Meta-Philosophical Reflection on Feminist Philosophies of Science*. Cham: Springer, pp. 75–89.

John, S. (2021). *Objectivity in Science.* Elements in the Philosophy of Science. Cambridge: Cambridge University Press.

Johnson, G. M. (2022). Excerpt from are algorithms value-free? Feminist theoretical virtues in machine learning. In K. Martin, ed., *Ethics of Data and Analytics*. New York: Auerbach, pp. 27–35.

Jukola, S. (2021). Commercial interests, agenda setting, and the epistemic trustworthiness of nutrition science. *Synthese, 198*(Suppl 10), 2629–46.

Kitcher, P. (2001). *Science, Truth, and Democracy.* Oxford: Oxford University Press.

Knox, B. (2022). Exclusion of the psychopathologised and hermeneutical ignorance threaten objectivity. *Philosophy, Psychiatry, & Psychology, 29*(4), 253–66.

Koertge, N. (2000). Science, values, and the value of science. *Philosophy of Science, 67* (Proceedings of the 1998 Biennial Meeting of the Philosophy of Science Association, Part II: Symposia Papers), 45–57.

Kourany, J. A. (2003). A philosophy of science for the twenty-first century. *Philosophy of science, 70*(1), 1–14.

Kourany, J. A. (2008). Replacing the ideal of value-free science. In M. Carrier, D. Howard, and J. Kourany, eds., *The Challenge of the Social and the Pressure of Practice: Science and Values Revisited*. Pittsburgh: University of Pittsburgh Press, 87–111.

Kourany, J. A. (2010). *Philosophy of Science after Feminism*. Oxford: Oxford University Press.

Kourany, J. A. (2016). Should some knowledge be forbidden? The case of cognitive differences research. *Philosophy of Science, 83*(5), 779–90.

Kreitzer, R. J., & Sweet-Cushman, J. (2021). Evaluating student evaluations of teaching: A review of measurement and equity bias in SETs and recommendations for ethical reform. *Journal of Academic Ethics, 20*, 1–12.

Kuhn, T. S. (1977). Objectivity, value judgement, and theory choice. In T. S. Kuhn, ed., *The Essential Tension: Selected Studies in Scientific Tradition and Change*. Chicago: Chicago University Press, pp. 320–39.

Kwon, D. (2022). The rise of citational justice: How scholars are making references fairer. *Nature, 603*(7902), 568–71.

Lee, C. J., Sugimoto, C. R., Zhang, G., & Cronin, B. (2013). Bias in peer review. *Journal of the American Society for Information Science and Technology, 64*(1), 2–17.

Leslie, S. J., Cimpian, A., Meyer, M., & Freeland, E. (2015). Expectations of brilliance underlie gender distributions across academic disciplines. *Science*, *347*(6219), 262–65.

Leuschner, A., & Fernández Pinto, M. (2021). How dissent on gender bias in academia affects science and society: Learning from the case of climate change denial. *Philosophy of Science*, *88*(4), 573–93.

Lloyd, E. A. (1995a). Objectivity and the double standard for feminist epistemologies. *Synthese*, *104*, 351–81.

Lloyd, E. A. (1995b). Feminism as method: What scientists get that philosophers don't. *Philosophical Topics*, *23*(2), 189–220.

Lloyd, E. A. (1999). Evolutionary psychology: The burdens of proof. *Biology and Philosophy*, *14*(2), 211–33.

Lloyd, E. A. (2005). *The Case of the Female Orgasm: Bias in the Science of Evolution*. Cambridge, MA: Harvard University Press.

Lloyd, G. (1984). *The Man of Reason: 'Male' and 'Female' in Western Philosophy*. London: Methuen.

Longino, H. E. (1990). *Science as Social Knowledge*. Princeton: Princeton University Press.

Longino, H. E. (1996). Cognitive and non-cognitive values in science: Rethinking the dichotomy. In L. H. Nelson and J. Nelson, eds., *Feminism, Science, and the Philosophy of Science*. Dordrecht: Springer, pp. 39–58.

Longino, H. E. (2002). *The Fate of Knowledge*. Princeton: Princeton University Press.

Longino, H. E. (2008). Values, heuristics, and the politics of knowledge. In M. Carrier, D. Howard, and J. Kourany, eds., *The Challenge of the Social and the Pressure of Practice: Science and Values Revisited*. Pittsburgh: Pittsburgh University Press, pp. 68–86.

Longino, H. E., & Doell, R. (1983). Body, bias, and behavior: A comparative analysis of reasoning in two areas of biological science. *Signs: Journal of Women in Culture and Society*, *9*(2), 206–27.

Marçal, K. (2021). *Mother of Invention: How Good Ideas Get Ignored in an Economy Built for Men*. London: Harper Collins.

Martin, E. (1991). The egg and the sperm: How science has constructed a romance based on stereotypical male-female roles. *Signs: Journal of Women in Culture and Society*, *16*(3), 485–501.

Maserejian, N. N., Link, C. L., Lutfey, K. L., Marceau, L. D., McKinlay, J. B. (2009). Disparities in physicians' interpretations of heart disease symptoms by patient gender: Results of a video vignette factorial experiment. *Journal of Women's Health*, *18*(10), 1661–67.

McMullin, E. (1983). Values in science. In P. D. Asquith and T. Nickles, eds., *PSA 1982 Volume II, Proceedings of the 1982 Biennial Meeting of the Philosophy of Science Association: Symposia*. East Lansing, 3–28.

Mengel, F., Sauermann, J., & Zölitz, U. (2019). Gender bias in teaching evaluations. *Journal of the European Economic Association*, *17*(2), 535–66.

Merchant, C. (1980). *The Death of Nature: Women, Ecology, and the Scientific Revolution*. San Francisco: Harper Collins.

Meynell, L. (2020). What's Wrong with (Narrow) Evolutionary Psychology. In S. Crasnow and K. Intemann, eds., *The Routledge Handbook of Feminist Philosophy of Science*. New York: Routledge, pp. 303–15.

Mikkola, M. (2023). Feminist perspectives on sex and gender. In E. N. Zalta, ed., *The Stanford Encyclopedia of Philosophy*, https://plato.stanford.edu/archives/spr2023/entries/feminism-gender/.

Mills, C. W. (2007). White ignorance. In S. Sullivan and N. Tuana, eds., *Race and Epistemologies of Ignorance*. New York: Suny Press, pp. 11–38.

MIT, Committee on Women Faculty in the School of Science (1999). A Study on the Status of Women Faculty in Science at MIT. *The MIT Faculty Newsletter*, 11(4).

Möbius, P. J. (1900/1907). *Ueber den physiologischen Schwachsinn des Weibes*. Halle a. d. S.: Marhold.

Nelson, L. H. (1990). *Who Knows: From Quine to a Feminist Empiricism*. Philadelphia: Temple University Press.

Nelson, L. H. (2020). Feminist philosophy of biology. In S. Crasnow and K. Intemann, eds., *The Routledge Handbook of Feminist Philosophy of Science*. New York: Routledge, pp. 263–74.

Nelson, L. H. (2022). Underdetermination, holism, and feminist philosophy of science. *Synthese*, *200*(1), 50.

Nezhat, C., Nezhat, F., & Nezhat, C. H. (2020). Endometriosis: Ancient disease, ancient treatments. In C. H. Nezhat, ed., *Endometriosis in Adolescents*. Cham: Springer, pp. 13–127.

NSB (National Science Board, National Science Foundation) (2023). Higher Education in Science and Engineering. *Science and Engineering Indicators 2024*. NSB-2023-32. Alexandria, VA. https://ncses.nsf.gov/pubs/nsb202332/.

Nussbaum, M. C. (2022). On not hating the body. *Liberties: A Journal of Culture and Politics*, *2*(2). https://libertiesjournal.com/articles/on-not-hating-the-body/.

Okruhlik, K. (1994). Gender and the biological sciences. *Canadian Journal of Philosophy*, *24*(sup1), 21–42.

Pinnick, C. L. (1994). Feminist epistemology: Implications for philosophy of science. *Philosophy of Science*, *61*(4), 646–57.

Pinnick, C. L., Koertge, N., & Almeder, R. F., eds. (2003). *Scrutinising Feminist Epistemology: An Examination of Gender in Science*. New Brunswick: Rutgers University Press.

Pohlhaus Jr., G. (2012). Relational knowing and epistemic injustice: Toward a theory of willful hermeneutical ignorance. *Hypatia*, *27*(4), 715–35.

Proctor, R. N., & Schiebinger, L. (2008). *Agnotology: The Making and Unmaking of Ignorance*. Stanford: Stanford University Press.

Quine, W. v. O (1953). Two dogmas of empiricism. In W. v. O. Quine, *From a Logical Point of View*, Cambridge, MA: Harvard University Press, pp. 20–46.

Quine, W. v. O. (1975). On empirically equivalent systems of the world. *Erkenntnis*, 9(3), 313–328.

Richardson, S. S. (2010). Feminist philosophy of science: History, contributions, and challenges. *Synthese*, *177*, 337–362.

Rolin, K. (2002). Gender and trust in science. *Hypatia*, *17*(4), 95–118.

Rolin, K. (2006). The bias paradox in feminist standpoint epistemology. *Episteme*, *3*(1-2), 125–136.

Rolin, K. (2017). Can Social Diversity be Best Incorporated into Science by Adopting the Social Value Management Ideal? In K. C. Elliott and D. Steel, eds., *Current Controversies in Values and Science*. New York: Routledge, pp. 113–29.

Rooney, P. (1991). Gendered reason: Sex metaphor and conceptions of reason. *Hypatia*, *6*(2), 77–103.

Rooney, P. (1992). On values in science: Is the epistemic/non-epistemic distinction useful?. In Grasswick, ed., *PSA: Proceedings of the Biennial Meeting of the Philosophy of Science Association* (Vol. 1992 no. 1). Cambridge, MA: Cambridge University Press., pp. 13–22.

Rooney, P. (2011). The marginalization of feminist epistemology and what that reveals about epistemology 'proper'. In H. Grasswick (ed.), *Feminist Epistemology and Philosophy of Science*. Dordrecht: Springer, 3–24.

Rooney, P. (2012). What is distinctive about feminist epistemology at 25. In S. L. Crasnow and A. M. Superson (eds.), *Out from the Shadows: Analytical Feminist Contributions to Traditional Philosophy*. Oxford: Oxford University Press, 339–375.

Rooney, P. (2017). The Borderlands between Epistemic and Non-Epistemic Values. In K. C. Elliott and D. Steel, eds., *Current Controversies in Values and Science*. New York: Routledge, pp. 31–45.

Sackett, D. L. (1979). Bias in analytic research. *Journal of Chronic Disease*, 32, 51–63.

Samulowitz, A., Gremyr, I., Eriksson, E., & Hensing, G. (2018). 'Brave men' and 'emotional women': A theory-guided literature review on gender bias in health care and gendered norms towards patients with chronic pain. *Pain Research and Management*, Article ID 6358624. www.hindawi.com/journals/prm/2018/6358624/.

Scheeman; N. (2001). Epistemology resuscitated: Objectivity as trustworthiness. In: N. Tuana & S. Morgen (eds.), *Engendering Rationalities*. Albany: State University of New York Press, pp. 23–52.

Schiebinger, L. (1999). *Has Feminism Changed Science?* Cambridge, MA: Harvard University Press.

Soble, A. (1995). In defense of Bacon. *Philosophy of the Social Sciences*, 25(2), 192–215.

Solomon, M. (2001). *Social Empiricism*. Cambridge, MA: The MIT Press.

Solomon, M. (2012). The web of valief: An assessment of feminist radical empiricism. In S. L. Crasnow and A. M. Superson, eds., *Out from the Shadows: Analytical Feminist Contributions to Traditional Philosophy*. New York: Oxford University Press, pp. 435–50.

Stanford, P. K. (2006). *Exceeding our Grasp: Science, History, and the Problem of Unconceived Alternatives*. Oxford: Oxford University Press.

Sugimoto, C. R., & Larivière, V. (2023). *Equity for Women in Science: Dismantling Systemic Barriers to Advancement*. Cambridge, MA: Harvard University Press.

Summers, L. (2005). Full transcript: Remarks at the NBER conference on diversity in the science and engineering workforce. *The Harvard Crimson* (18.02.2005). https://www.thecrimson.com/article/2005/2/18/full-transcript-president-summers-remarks-at/.

Tasca, C., Rapetti, M., Carta, M. G., & Fadda, B. (2012). Women and hysteria in the history of mental health. *Clinical Practice and Epidemiology in Mental Health*, *8*, 110–19.

Thornhill, R., & Palmer, C. (2000). *A Natural History of Rape*. Cambridge, MA: MIT Press.

Travis, C. B. (ed.). (2003). *Evolution, Gender, and Rape*. Cambridge, MA: MIT Press.

Tremain, S. L. (2020). Naturalising and denaturalising impairment and disability in philosophy and feminist philosophy of science. In S. Crasnow and K. Intemann, eds., *The Routledge Handbook of Feminist Philosophy of Science*. New York: Routledge, pp. 144–156.

Trivers, Robert (1972). Parental Investment and Sexual Selection. In B. Campbell (ed.), Sexual Selection and the Descent of Man: 1871–1971. Chicago: Aldine, 136–179.

Tuana, N. (1993). *The Less Noble Sex: Scientific, Religious, and Philosophical Conceptions of Woman's Nature*. Bloomington: Indiana University Press.

Tuana, N. (2004). Coming to understand: Orgasm and the epistemology of ignorance. *Hypatia, 19*(1), 194–232.

Tuana, N. (2006). The speculum of ignorance: The women's health movement and epistemologies of ignorance. *Hypatia, 21*(3), 1–19.

Urry, C. M. (2008). Are Photons gendered? Women in Physic and Astronomy. In L. Schiebinger (ed.), *Gendered Innovations in Science and Engineering*. Stanford: Stanford University Press, 150–164.

Wang, M. T., Degol, J. L. (2017). Gender Gap in Science, Technology, Engineering, and Mathematics (STEM): Current Knowledge, Implications for Practice, Policy, and Future Directions. *Educational Psychology Review* **29**, 119–140.

Whyte, K. P., & Crease, R. P. (2010). Trust, expertise, and the philosophy of science. *Synthese, 177*, 411–425.

Williams, W. M., & Ceci, S. J. (2015). National hiring experiments reveal 2:1 faculty preference for women on STEM tenure track. *PNAS 112*(17), 5360–65.

Wylie, A. (2003). Why standpoint matters. In R. Figueroa and S. Harding, eds., *Science and other Cultures: Issues in Philosophies of Science and Technology*. New York: Routledge, pp. 26–48.

Wylie, A. (2012). Feminist philosophy of science: Standpoint matters. *Proceedings and Addresses of the American Philosophical Association* (Vol. 86, no. 2). Newark: American Philosophical Association, pp. 47–76.

Yap, A. (2016). Feminist radical empiricism, values, and evidence. *Hypatia, 31*(1), 58–73.

Acknowledgements

I first encountered feminist philosophy of science as a Master's student roughly fifteen years ago. This discovery started an academic journey that allowed me to meet a lot of great people and impressive scholars. While it is impossible to list them all here, I am profoundly grateful to everyone who has helped, taught, and inspired me along the way.

I am particularly indebted to Torsten Wilholt for his support and encouragement throughout my career. I also want to thank Saana Jukola and Somogy Varga for their insightful feedback on earlier versions of the manuscript; Lucy Seton-Watson for her assistance with the English language; and two anonymous reviewers for their careful reading and constructive critiques. Additionally, I want to thank the series editor, Jacob Stegenga, for his guidance and patience throughout the process.

Last but certainly not least, I want to express my gratitude to my family and friends. In particular, I am deeply thankful to Najko Jahn for his insight, love, and unwavering support.

Cambridge Elements

Philosophy of Science

Jacob Stegenga
University of Cambridge

Jacob Stegenga is a Reader in the Department of History and Philosophy of Science at the University of Cambridge. He has published widely on fundamental topics in reasoning and rationality and philosophical problems in medicine and biology. Prior to joining Cambridge he taught in the United States and Canada, and he received his PhD from the University of California San Diego.

About the Series

This series of Elements in Philosophy of Science provides an extensive overview of the themes, topics and debates which constitute the philosophy of science. Distinguished specialists provide an up-to-date summary of the results of current research on their topics, as well as offering their own take on those topics and drawing original conclusions.

Cambridge Elements

Philosophy of Science

Elements in the Series

Fundamentality and Grounding
Kerry McKenzie

Values in Science
Kevin C. Elliott

Scientific Representation
James Nguyen and Roman Frigg

Philosophy of Open Science
Sabina Leonelli

Natural Kinds
Muhammad Ali Khalidi

Scientific Progress
Darrell P. Rowbottom

Modelling Scientific Communities
Cailin O'Connor

Logical Empiricism as Scientific Philosophy
Alan W. Richardson

Scientific Models and Decision Making
Eric Winsberg and Stephanie Harvard

Science and the Public
Angela Potochnik

Abductive Reasoning in Science
Finnur Dellsén

Feminist Philosophy of Science
Anke Bueter

A full series listing is available at: www.cambridge.org/EPSC

Printed in the United States
by Baker & Taylor Publisher Services